ELECTRICAL DESIGN
of Commercial and Industrial Buildings

John Hauck
Instructor, Electrical Technology
Long Beach City College
Long Beach, California

JONES AND BARTLETT PUBLISHERS
Sudbury, Massachusetts
BOSTON   TORONTO   LONDON   SINGAPORE

World Headquarters
Jones and Bartlett Publishers
40 Tall Pine Drive
Sudbury, MA 01776
978-443-5000
info@jbpub.com
www.jbpub.com

Jones and Bartlett Publishers Canada
6339 Ormindale Way
Mississauga, Ontario L5V 1J2
Canada

Jones and Bartlett Publishers International
Barb House, Barb Mews
London W6 7PA
United Kingdom

Jones and Bartlett's books and products are available through most bookstores and online booksellers. To contact Jones and Bartlett Publishers directly, call 800-832-0034, fax 978-443-8000, or visit our website, www.jbpub.com.

Substantial discounts on bulk quantities of Jones and Bartlett's publications are available to corporations, professional associations, and other qualified organizations. For details and specific discount information, contact the special sales department at Jones and Bartlett via the above contact information or send an email to specialsales@jbpub.com.

Production Credits

Chief Executive Officer: Clayton Jones
Chief Operating Officer: Don W. Jones, Jr.
President, Higher Education and Professional Publishing: Robert W. Holland, Jr.
V.P., Sales: William J. Kane
V.P., Design and Production: Anne Spencer
V.P., Manufacturing and Inventory Control: Therese Connell
Publisher: Kimberly Brophy
Acquisitions Editor—Electrical: Martin Schumacher
Associate Managing Editor: Robyn Schafer
Production Manager: Jenny L. Corriveau
Production Assistant: Tina Chen
Associate Marketing Manager: Meagan Norlund
Assistant Photo Researcher: Carolyn Arcabascio
Composition: Modern Graphics, Inc.
Cover Design: Kristin E. Parker
Cover Image: © Ulrich Mueller/Dreamstime.com
Printing and Binding: Courier Stoughton
Cover Printing: Courier Stoughton

Copyright © 2011 by Jones and Bartlett Publishers, LLC

All rights reserved. No part of the material protected by this copyright may be reproduced or utilized in any form, electronic or mechanical, including photocopying, recording, or by any information storage and retrieval system, without written permission from the copyright owner.

The procedures and protocols in this book are based on the most current recommendations of responsible sources. The publisher, however, makes no guarantee as to, and assumes no responsibility for, the correctness, sufficiency, or completeness of such information or recommendations. Other or additional safety measures may be required under particular circumstances.

Additional photographic and illustration credits appear on page 164, which constitutes a continuation of the copyright page.

Library of Congress Cataloging-in-Publication Data
Hauck, John.
 Electrical design of commercial and industrial buildings / John Hauck.
   p. cm.
 Includes index.
 ISBN 978-0-7637-5828-8
 1. Electric wiring, Interior. 2. Commercial buildings—Electric equipment. I. Title.
 TK3284.H38 2009
 621.319'24—dc22
                                2009024192

6048

Printed in the United States of America
13 12 11 10 09   10 9 8 7 6 5 4 3 2 1

CONTENTS

Chapter 1 Electrical Plan Design........................ 1
   Introduction 2
   The Design Process 2
   Understanding the Project Scope 2
   Defining Parts of the Electrical Plan 2
      General Electrical Requirements............................. 2
      Specialized Electrical Requirements 2
      Lighting Systems 3
      Distribution System 3
   Determining Applicable Standards 3
   Creating the Electrical Plan 4
   Wrap Up 7
   You Are the Designer 9

Chapter 2 General Electrical Requirements............ 11
   Introduction 12
   Determining General Purpose Requirements 12
   Determining Branch-Circuit Requirements........................ 12
      Balancing the Branch-Circuit Distribution 14
      Panel Schedules 15
   Designating Branch Circuits in the Electrical Plan 16
   Wrap Up 18
   You Are the Designer 22

Chapter 3 Specialized Electrical Requirements......... 27
   Introduction 28
   Equipment List 28
   Specialized Equipment Branch Circuits 28
   Motors 30
      Motor Equipment Branch-Circuit Conductors 31
      Motor Branch-Circuit Short-Circuit and
         Ground-Fault Protection Devices 34
      Motor Branch-Circuit Overload Devices 34
      Grounding of Motors and Equipment......................... 36
   Raceways for Branch-Circuit Distribution 38

Determining the Number of Panelboards	**40**
Larger Motors	42
Equipment Branch Circuits and Panel Schedules	42
Raceway Legends	**43**
Wrap Up	**44**
You Are the Designer	**48**

Chapter 4   Lighting Systems 55

Introduction	**56**
Determining Lighting Requirements	**56**
Selecting Lighting Fixtures	**57**
Calculating the Number of Lighting Fixtures Required	**57**
Determining Lighting Fixture Location	**59**
Creating the Lighting Plan	**59**
Lighting Fixture Schedule	59
Branch-Circuit Identification	62
Control Devices	63
Lighting Branch-Circuit Raceways	65
Including Keynotes Sections	**65**
Complying with Energy Code Requirements	66
Interpreting Energy Code Design Methods	66
Energy Management	67
Wrap Up	**69**
You Are the Designer	**73**

Chapter 5   Distribution Systems 85

Introduction	**86**
Switchboards	86
Single-Line Diagrams	**86**
Service Conductors	88
Distribution	88
Panelboard, Transformer, and Branch-Circuit Distributions	89
Distribution Transformers	89
System Grounding Methods	89
Selecting and Sizing Distribution System Components	90
Panelboards and Feeders	91
Distribution Transformers	91

System Grounding 91
　　　　　Grounding Distribution Transformers 94
　　　　　Grounding Service Entrance Equipment 95
　　Using Conductors in Parallel 95
　　Wrap Up 97
　　You Are the Designer 100

Chapter 6 Load and Short-Circuit Calculations 109

　　Introduction 110
　　Performing Load Calculations 110
　　　　　General Electrical Loads 110
　　　　　Specialized Electrical Loads 111
　　　　　Lighting System Loads 111
　　　　　Distribution System Loads 111
　　Performing Short-Circuit Calculations 112
　　　　　Transformer Impedance 113
　　Wrap Up 119
　　You Are the Designer 122

Chapter 7 Electrical Plan Review 127

　　Introduction 128
　　Electrical Plan Review Process 128
　　Electrical Plan Review Checklist 128
　　　　　General and Specialized Electrical Requirements 129
　　　　　Lighting System 129
　　　　　Distribution System 129
　　Wrap Up 131
　　You Are the Designer 133

Glossary 137

Appendix 143

Index 161

Photo Credits 164

CHAPTER RESOURCES

Electrical Design of Commercial and Industrial Buildings provides students with the ability to understand how electrical plans are conceptualized, designed, and compiled, and illustrates how to apply this knowledge to the projects they will encounter in the field. The unique hands-on approach of this text does much more than simply relay theory and teach the ability to read electrical plans; it provides real-world examples, relevant codes and requirements, and an opportunity to increase student knowledge about key aspects of an electrical design. Features that reinforce and expand on essential information include:

TIPS
Helpful tips offer additional advice or reinforce information provided in the text.

CALCULATION EXAMPLES
Step-by-step examples guide the reader through essential calculations.

YOU ARE THE DESIGNER
This unique hands-on student project integrates knowledge presented in the text and guides students through the creation of their own electrical design project.
- **About Your Project**
- **Resources**
- **Get to Work:** In this section students will apply the concepts presented in the chapter to the creation of their own electrical design plan.

CHAPTER RESOURCES

Wrap Up

■ Master Concepts

- Equipment lists outlining all the equipment in a facility are necessary to determine the number of branch circuits needed and the quantity and operating voltages of panelboards.
- Because of their amperage requirements and manufacturer specifications, specialized equipment and motorized equipment must be served by their own individual branch circuits.
- When loads are located farther than 100 ft from their source, because of the resistance of the conductors serving specialized and motorized equipment loads, the circular mil area of branch circuit conductors must be increased to provide for a maximum voltage drop of 3 percent of the source voltage.
- Properly determining the size of the overcurrent protection for motors, equipment, and the branch circuit conductors that serve them is very important to protect these devices from short-circuit or ground-fault conditions.
- When designing panelboards, designers should aim to meet all the minimum requirements and provide an efficient system to serve the facility safely with an ability to expand for future needs.
- A raceway legend, though not a requirement, is an extremely useful tool to identify the raceways and the conductors contained within them for identification on a design plan.

■ Charged Terms

277/480-volt, 3-phase, 4-wire, wye system A distribution system generated with three individual sine waves separated by 120 electrical degrees that are identified as phases A, B, and C. One leg of each of the three phase coils is electrically connected to the others at a common point, forming a wye, which when grounded becomes the fourth wire (or neutral) in the system. This allows for each of the three individual phase voltages to supply 277 volts to the grounded point, while the line voltage across each of the phases produces 480 volts. The line-to-line voltages can supply both 480-volt 3-phase and 480-volt single-phase. This system is commonly used in commercial applications where 480 volts is required for machinery loads and in applications to serve 277-volt lighting loads.

equipment grounding conductor The conductive path installed to connect normally non-current-carrying metal parts of equipment together and to the system grounded conductor, to the grounding electrode conductor, or to both [100].

equipment list A developed table that lists details about the specialized equipment that is to be incorporated into a design plan.

ground fault A condition in which high levels of current could flow when an ungrounded conductor accidentally comes in contact with a grounded reference.

inrush current A momentary high level of amperage flowing in a circuit such as those associated with motorized equipment loads.

...grounding conductor that, when installed...separate from a grounding method...ed for electrically sensitive equipment...ies. When isolated grounding is provided through the use of receptacles, the receptacle must be identified on the design plan as "IG."

manufacturer electrical specification sheet Information provided by a manufacturer that lists specific details about the product; these specification sheets are often used to obtain information about motorized equipment and lighting fixtures.

raceway legend A table developed by an electrical designer that illustrates information about raceways installed for a project.

short circuit A dangerous condition in which circuit conductors contact each other and reduce the intended ohmic resistance of the circuit; often referred to as a line-to-line or line-to-neutral short. (*See also* ground fault.)

voltage drop A loss of voltage on a conductor resulting from the length of the conductor, its resistance, and the amperage imposed on the conductor.

■ Check Your Knowledge

1. The purpose of an equipment list is to:
   A. list the electrical requirements for any equipment to be installed on a project.
   B. provide manufacturer specification sheets to the owner.
   C. assign each equipment piece an identification number.
   D. assign each equipment piece a branch circuit number.
2. A raceway legend provides:
   A. the types and quantities of raceway material to be used on a project.
   B. a detailed layout for the routing of all the raceways for a project.
   C. a list of raceways by type, size, and the conductors to serve equipment loads.
   D. a listing of all the raceway types for a project.
3. A 10-hp, 208-volt, 3-phase motor located 50 ft from the source requires a size _____ AWG copper Type THWN conductor.
   A. 12
   B. 10
   C. 8
   D. 6

WRAP UP
End-of-chapter materials and activities reinforce important concepts and evaluate student comprehension.
- **Master Concepts** thoroughly summarize chapter concepts.
- **Charged Terms** provide key terms and definitions from the chapter.
- **Check Your Knowledge** promotes critical thinking and tests students' understanding of key concepts.

STUDENT AND INSTRUCTOR RESOURCES

Student Resource CD-ROM
Packaged with each copy of *Electrical Design of Commercial and Industrial Buildings*, this CD-ROM provides students with valuable resources to complete their own electrical design projects. It includes completed and partially-completed electrical designs in addition to many valuable reference materials such as an electrical symbol list, sample lighting fixture schedule, and electrical plan review checklist.

Instructor's ToolKit CD-ROM
We've made it easy for you with a fully adaptable Instructor's ToolKit CD-ROM, featuring:
- **Lecture Outlines**—complete, ready-to-use lesson plans that outline each of the chapter's major topics
- **PowerPoint™ Presentations**—thorough and engaging presentations for each of the book's chapters
- **ImageBank**—all of the images found in the text, clearly organized for easy access
- **Answer Key**—for all the questions found in the text, clearly organized by chapter

ACKNOWLEDGEMENTS

The author and Jones and Bartlett Publishers would like to thank the following individuals for their thoughtful reviews of this text and valuable feedback:

Stanley R. Flippin
Director of Education, IEC Dallas
Dallas, Texas

Ron T. Murray
Master Electrician, Lincoln Technical Institute
New Britain, Connecticut

Kenneth W. Mondon
Electrical Instructor, Vatterott College
Licensed Master Electrician, Mondon Electric
Quincy, Illinois

Alan W. Stanfield
Electrical Instructor and Contractor,
Griffin Technical College
Griffin, Georgia

DEDICATION

Thank you to all my family members and colleagues—without your support, this text would not have been written.

A special thank you to Kim Anderson, for without your knowledge and guidance on the curriculum for this text and so many other projects over the years, this book never would have been possible.

Also, a thank you to Marty, Robyn, and Tina at Jones and Bartlett Publishers. Combined, their expertise, talents, and creativity have allowed for ideas to translate to the written page.

INTRODUCTION

In my experience as both an electrical contractor and educator, I have found that there are many highly skilled technicians in the electrical industry, and those with less experience, who wish to enroll in a well-taught course in reading electrical design plans. The students in these courses may become very skilled in the interpretation of the electrical symbols on electrical designs, but they often complete the course without a full understanding of how the plan was conceptualized, designed, and compiled.

Without this full understanding of how electrical prints are developed (including all the design criteria, electrical codes, requirements, and so forth), the technician is limited in his or her ability to make additions or changes to an electrical design in the field, as is often necessary. In construction, a contract between the electrical contractor and the owner (or owner's representative) will often specify that any changes to the electrical wiring system that occur during the construction phase of the project shall be performed and documented on a set of "as-built" drawings which should be provided to the owner by the electrical contractor. If the electrical technician does not have the ability to produce these drawings due to a lack of understanding, then the drawings must be produced by an outside source. This will most likely lead to additional costs to the electrical contractor and possible delays in the completion of the project.

For those working in electrical maintenance positions within facilities, the same is true. When any changes, additions, or deletions are made to the electrical system or wiring within the equipment, documentation is required to maintain current records. This allows for those performing electrical maintenance procedures to have current information for all the electrical sources serving the facilities and the machinery, thus providing for the greatest degree of electrical safety.

This text is the first of its kind to provide technicians with the knowledge they need to not only read electrical design plans, but to understand electrical wiring design. This text is not intended to be an introduction to electrical print reading; its goal is to expand upon existing knowledge, presenting all the required components of the electrical design and serving as a good reference for those students who are, or will be, working on an electrical design.

For those who wish to have the ability to create their own design plans, the unique "You Are the Designer" feature at the end of each chapter assists readers in applying the concepts and ideas to their own design. Through its unique hands-on approach, this text uses completed and partially completed designs to guide students through the creation of their own electrical design specifications, applying all the regulations as required by the *National Electrical Code*® (*NEC*®) and incorporating all the necessary industry standards. The electrical plans are provided in hardcopy in the text and on the *Student Resource CD-ROM*, for those who have access and experience with AutoCAD and wish to complete the project in a computer based format.

John Hauck
Electrical Technology
Long Beach City College
Long Beach, California

CHAPTER 1

Electrical Plan Design

Chapter Outline

- Introduction
- The Design Process
- Understanding the Project Scope
- Defining Parts of the Electrical Plan
- Determining Applicable Standards
- Creating the Electrical Plan

Learning Objectives

- Identify the steps in the electrical design process.
- Determine the scope of an electrical design project.
- Interpret the various components of an electrical plan, including general and specialized loads, lighting systems, and distribution systems.
- Recognize the symbols used in electrical plan design.
- Identify the standards and regulations that guide the electrical design process.

Introduction

For all building construction or remodeling building projects, the owner or occupant must first have a concept for the new design, and then the architect or designer can produce a set of building plans. These plans convey all the required information to the local inspection authority and associated building trades so that the construction or remodeling can take place. Because commercial and industrial buildings contain a number of electrical systems, these plans include specific electrical designs and additional documentation to verify that the design conforms to all required building codes.

The Design Process

An electrical design goes through several important stages of development. First, the designer must understand the scope of the project. Then, the designer defines and designs each component (such as general office areas, specialized machinery, and power distribution equipment) to recognized industry standards. Finally, these individual components are compiled to form the final presentation for the design.

Understanding the Project Scope

Every electrical design has unique requirements, depending on the scope of the project. The project scope is determined by the customer's requirements and the type of structure that the customer will occupy. For example, if the project requires new electrical systems for an existing building, then the electrical designer works to incorporate all the new electrical wiring into the existing system. The designer must evaluate the existing electrical system to ensure that existing electrical systems can accommodate new additional electrical loads that will be imposed on them. When the design is for a new proposed facility, then the scope of the project is much greater. Electrical designs for these types of projects require an entirely new electrical system design.

Defining Parts of the Electrical Plan

Depending on the overall scope of the project, a design can include the following components:
- General electrical requirements (e.g., general purpose receptacles)
- Specialized electrical requirements (e.g., specialized office equipment or machinery)
- Lighting systems
- Electrical distribution systems

General Electrical Requirements

General electrical requirements should be defined first on any electrical design project. General electrical requirements are items such as the 120-volt general purpose receptacle outlets located throughout the commercial or industrial building. These receptacles are usually not specified to serve any particular load but rather are for general purpose use such as for desktop devices, standard wall receptacles, and desktop computer equipment with no special electrical requirements.

Specialized Electrical Requirements

Certain projects may include specialized electrical equipment that requires separate or dedicated electrical circuitry that serves only the specialized equipment (see **FIGURE 1-1**). This equipment may be of the following types:
- Computers and/or network servers
- Photocopiers
- Microwave ovens and other lunchroom appliances
- Vending machines
- Machinery and equipment

Because of their electrical load requirements, as per the manufacturer's requirements, these pieces of equipment may require individual circuitry and special grounding methods (see Chapter 3).

FIGURE 1-1 Some commercial electrical equipment may have specialized electrical requirements.

Lighting Systems

Because of their complexity, lighting systems are the part of the design process that generally requires the greatest amount of time to develop (see **FIGURE 1-2**). These systems include all the lighting fixtures and their controls. Lighting systems have very detailed requirements as per the *NEC* and mandated energy requirements which require documentation showing that the system incorporates all required energy-saving technologies (see Chapter 4).

Distribution System

An electrical distribution system is the installed equipment that provides for the distribution of electrical wiring throughout the facility (see **FIGURE 1-3**). It includes the main switchboard, which receives the power source from the serving utility, and all the associated components such as panelboards that distribute all the required branch circuits throughout the facility (see Chapter 5). Part of the process of designing the distribution system is calculating the facility's amperage load and short-circuit val-

FIGURE 1-3 An electrical distribution system provides power to the entire facility.

ues; these calculations determine the total electrical demand requirements of the facility based on the individual parts of the electrical distribution system (see Chapter 6).

■ Determining Applicable Standards

Once each part of the design plan has been defined, the next stage is to design each part to industry-recognized standards as well as any additional standards set forth by the local jurisdiction for commercial or industrial occupancies. The primary industry standard is the **National Electrical Code (NEC)**, published by the National Fire Protection Association (NFPA). The *NEC* (commonly referred to as "the *Code*") is revised every three years and results in the publication of a new edition (e.g., the 2005 *NEC* or the 2008 *NEC*). Although the *Code* is applied on a national level, some local jurisdictions may have additional standards that exceed the requirements of the *NEC* or they may use a previous edition of the *Code*.

FIGURE 1-2 Lighting systems are the most complex part of an electrical design.

Always contact local building inspection authorities during the design process to determine whether any local standards have been adopted and which edition of the *NEC* is being enforced.

For projects based on a national template (such as is often the case with retail outlets and fast-food chains), any requirements or adjustments that are necessary to conform to local code requirements should be documented in the final plan in a general notes section. Please note that only officially documented standards may be enforced, not widespread, unofficial community practices.

Some projects will also have additional requirements based on their specific components, such as those including specialized electrical equipment. An electrical designer should always consider manufacturer guidelines for specialized equipment and use the appropriate electrical equipment standards set forth by the manufacturer for overcurrent protection sizes, specialized grounding requirements, and so forth. These specialized requirements may require that additional specialized wiring practices be observed; when this is the case, these specialized requirements must be documented on the plan.

Designers must also consider the standards of the **National Electrical Manufacturers Association (NEMA)**, which includes standards for motor lead identification, transformer terminal markings, plug and receptacle devices, and amperage ratings, and the **Electrical Apparatus and Service Association (EASA)**, which provides current and updated information for motors and controls. Designs that include lighting systems must conform to national or state-mandated energy-saving requirements. Designers should consult the **Illuminating and Engineering Society of North America (IESNA)** standards for lights and lighting products and properly document the design to ensure that it meets all the required criteria (see Chapter 4).

For projects that include new or upgraded parts of distribution systems served from a local utility, designers must consider any requirements set forth by the serving utility. These requirements may dictate the wiring methods and equipment required for the proper distribution from the serving utility to the customer (see Chapter 5). Calculated load values must reference manufacturer guidelines to ensure that distribution systems will support these loads (see Chapter 6).

In all cases, designers must have not only solid electrical knowledge and a thorough understanding of the electrical calculations and their necessity but also awareness of the application of all relevant codes and standards utilized within the electrical industry.

■ Creating the Electrical Plan

Once the various parts and applicable standards have been determined, the designer begins compiling those parts to form the electrical design and complete a set of plans.

Historically, these plans took the form of hand-drawn blueprints (see **FIGURE 1-4**), but today most plans are created digitally using **computer-aided design (CAD)** software tools. Digitized plans are easier to revise and transmit than those drawn with pen and pencil. When printed, digital plans are typically produced on standard-sized architectural plan sheets; the most common size sheets are architectural D sheets which are 24 in. × 36 in. and architectural E sheets which are 36 in. × 48 in.

On the plans, each device should be referenced using the appropriate electrical symbol. Electrical symbols allow for universal recognition of each part by the many people who will be working on the project so that they can estimate costs appropriately and construct the project to the specifications. The standardized electrical symbols used for building plans are provided by the **American National Standards Institute (ANSI)** (see **FIGURE 1-5**).

Not all symbols are used on every project, so the specific symbols used on a particular project

FIGURE 1-4 Traditional hand-drawn blueprints are no longer common because most designers use CAD design tools.

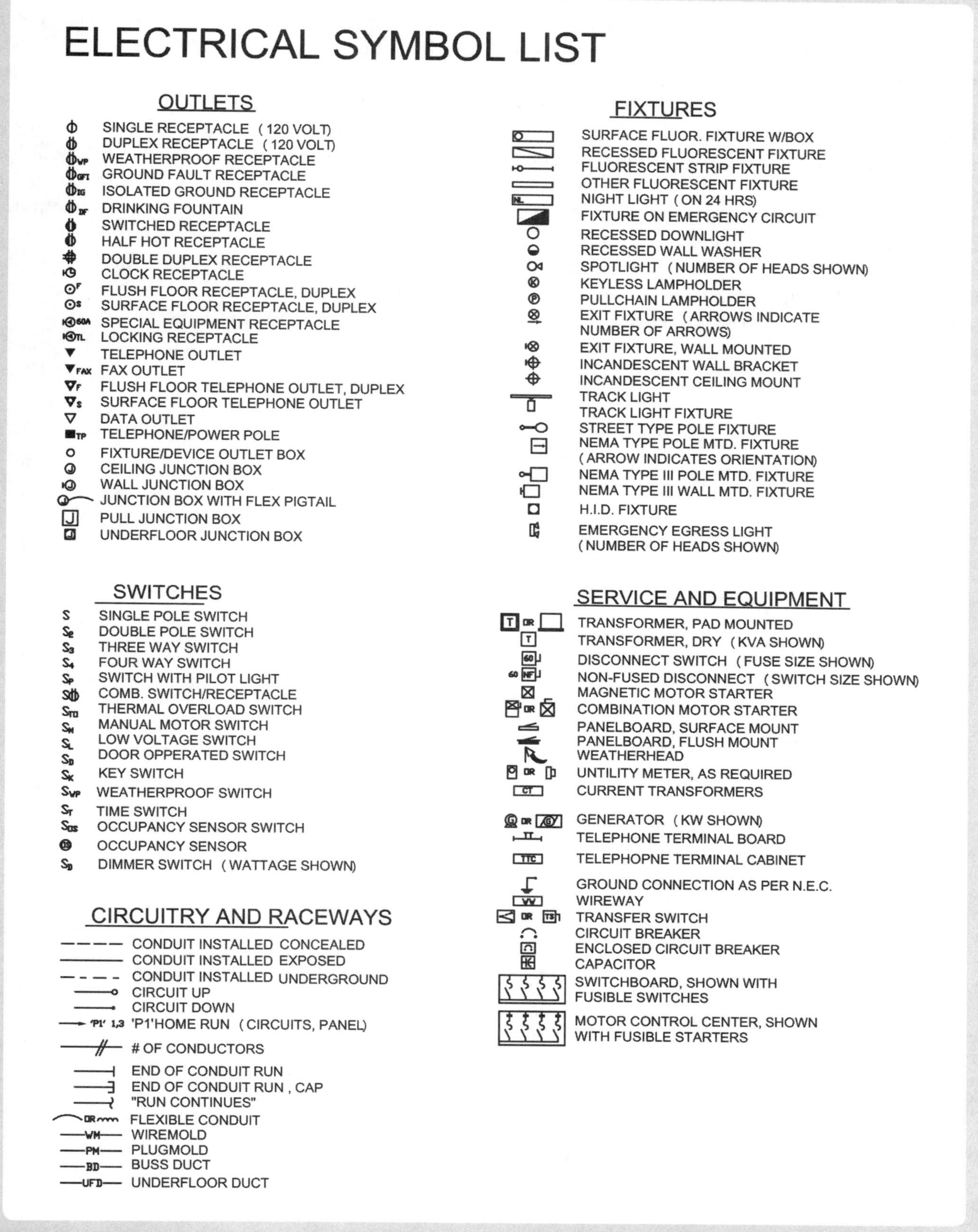

ELECTRICAL SYMBOL LIST

OUTLETS

- SINGLE RECEPTACLE (120 VOLT)
- DUPLEX RECEPTACLE (120 VOLT)
- WEATHERPROOF RECEPTACLE
- GROUND FAULT RECEPTACLE
- ISOLATED GROUND RECEPTACLE
- DRINKING FOUNTAIN
- SWITCHED RECEPTACLE
- HALF HOT RECEPTACLE
- DOUBLE DUPLEX RECEPTACLE
- CLOCK RECEPTACLE
- FLUSH FLOOR RECEPTACLE, DUPLEX
- SURFACE FLOOR RECEPTACLE, DUPLEX
- SPECIAL EQUIPMENT RECEPTACLE
- LOCKING RECEPTACLE
- TELEPHONE OUTLET
- FAX OUTLET
- FLUSH FLOOR TELEPHONE OUTLET, DUPLEX
- SURFACE FLOOR TELEPHONE OUTLET
- DATA OUTLET
- TELEPHONE/POWER POLE
- FIXTURE/DEVICE OUTLET BOX
- CEILING JUNCTION BOX
- WALL JUNCTION BOX
- JUNCTION BOX WITH FLEX PIGTAIL
- PULL JUNCTION BOX
- UNDERFLOOR JUNCTION BOX

SWITCHES

- S   SINGLE POLE SWITCH
- S₂ DOUBLE POLE SWITCH
- S₃ THREE WAY SWITCH
- S₄ FOUR WAY SWITCH
- S_P SWITCH WITH PILOT LIGHT
- S↺ COMB. SWITCH/RECEPTACLE
- S_TO THERMAL OVERLOAD SWITCH
- S_M MANUAL MOTOR SWITCH
- S_L LOW VOLTAGE SWITCH
- S_D DOOR OPPERATED SWITCH
- S_K KEY SWITCH
- S_WP WEATHERPROOF SWITCH
- S_T TIME SWITCH
- S_OS OCCUPANCY SENSOR SWITCH
- OCCUPANCY SENSOR
- S_B DIMMER SWITCH (WATTAGE SHOWN)

CIRCUITRY AND RACEWAYS

- ---- CONDUIT INSTALLED CONCEALED
- ——— CONDUIT INSTALLED EXPOSED
- - - - CONDUIT INSTALLED UNDERGROUND
- CIRCUIT UP
- CIRCUIT DOWN
- 'P1' 1,3 'P1' HOME RUN (CIRCUITS, PANEL)
- // # OF CONDUCTORS
- END OF CONDUIT RUN
- END OF CONDUIT RUN, CAP
- "RUN CONTINUES"
- FLEXIBLE CONDUIT
- WM WIREMOLD
- PM PLUGMOLD
- BD BUSS DUCT
- UFD UNDERFLOOR DUCT

FIXTURES

- SURFACE FLUOR. FIXTURE W/BOX
- RECESSED FLUORESCENT FIXTURE
- FLUORESCENT STRIP FIXTURE
- OTHER FLUORESCENT FIXTURE
- NIGHT LIGHT (ON 24 HRS)
- FIXTURE ON EMERGENCY CIRCUIT
- RECESSED DOWNLIGHT
- RECESSED WALL WASHER
- SPOTLIGHT (NUMBER OF HEADS SHOWN)
- KEYLESS LAMPHOLDER
- PULLCHAIN LAMPHOLDER
- EXIT FIXTURE (ARROWS INDICATE NUMBER OF ARROWS)
- EXIT FIXTURE, WALL MOUNTED
- INCANDESCENT WALL BRACKET
- INCANDESCENT CEILING MOUNT
- TRACK LIGHT
- TRACK LIGHT FIXTURE
- STREET TYPE POLE FIXTURE
- NEMA TYPE POLE MTD. FIXTURE (ARROW INDICATES ORIENTATION)
- NEMA TYPE III POLE MTD. FIXTURE
- NEMA TYPE III WALL MTD. FIXTURE
- H.I.D. FIXTURE
- EMERGENCY EGRESS LIGHT (NUMBER OF HEADS SHOWN)

SERVICE AND EQUIPMENT

- TRANSFORMER, PAD MOUNTED
- TRANSFORMER, DRY (KVA SHOWN)
- DISCONNECT SWITCH (FUSE SIZE SHOWN)
- NON-FUSED DISCONNECT (SWITCH SIZE SHOWN)
- MAGNETIC MOTOR STARTER
- COMBINATION MOTOR STARTER
- PANELBOARD, SURFACE MOUNT
- PANELBOARD, FLUSH MOUNT
- WEATHERHEAD
- UNTILITY METER, AS REQUIRED
- CURRENT TRANSFORMERS
- GENERATOR (KW SHOWN)
- TELEPHONE TERMINAL BOARD
- TELEPHOPNE TERMINAL CABINET
- GROUND CONNECTION AS PER N.E.C.
- WIREWAY
- TRANSFER SWITCH
- CIRCUIT BREAKER
- ENCLOSED CIRCUIT BREAKER
- CAPACITOR
- SWITCHBOARD, SHOWN WITH FUSIBLE SWITCHES
- MOTOR CONTROL CENTER, SHOWN WITH FUSIBLE STARTERS

FIGURE 1-5 Electrical symbols are used to indicate the parts of an electrical design.

should be included in a symbols list and attached to the final design. Occasionally the need may arise for a symbol that has not been developed (such as a symbol for a newer energy-saving or energy management device). In this case, the designer may create a new symbol for the electrical design plan, as long as it is added to the symbols list included with the plan.

Electrical design plans may be included as a separate document within a complete set of building plans. To identify the electrical plans, each page of the electrical design plan is labeled and numbered: E_1, E_2, E_3, and so forth. Please note that these electrical sheets (often called "E sheets") are not *architectural* E sheets, which denote a standard size paper. Electrical sheets are generally presented in the following order:
- Exterior electrical site plan
- Interior electrical power plan
- Interior lighting plan
- Documentation (such as panel schedules, electrical calculations, single line diagrams, and lighting system energy requirements)

The number of electrical sheets required for a project varies based on the amount of information that each project requires and how much of that information can fit on one page and still provide for a clear, concise, and understandable set of prints.

Wrap Up

■ Master Concepts

- When designing an electrical plan, the designer must understand the scope of the project, define each required part, and then design them to recognized industry standards.
- The major parts of an electrical plan include general and specialized electrical requirements, lighting systems, and the electrical distribution system.
- Every electrical plan must be designed to recognized industry standards, use appropriate electrical symbols, and conform to all applicable codes.
- Once the various parts and applicable standards have been determined, the designer creates a complete set of plans.

■ Charged Terms

<u>American National Standards Institute (ANSI)</u> An organization that oversees the development of standards created by manufacturers throughout the industry to promote safety and other standards.

<u>computer-aided design (CAD)</u> The use of computers and design software to aid in the design of drawings, objects, shapes, and other items.

<u>Electrical Apparatus and Service Association (EASA)</u> An organization that provides information and education about sales, service, and maintenance materials for motors, generators, and other electromechanical equipment.

<u>Illuminating and Engineering Society of North America (IESNA)</u> An organization that works with manufacturers, designers, architects, consultants, electrical and building contractors, and suppliers with regard to lighting systems.

<u>National Electrical Code (NEC)</u> Regulatory code published by the National Fire Protection Association (NFPA); also known as NFPA 70.

<u>National Electrical Manufacturers Association (NEMA)</u> A trade association that provides standards for the electrical manufacturing industry including the generation, transmission and distribution, control, and end use of electricity.

■ Check Your Knowledge

1. Which of the following parts are included in an electrical design?
   - A. General and specialized equipment loads
   - B. Lighting systems
   - C. Distribution system
   - D. All of the above

2. The standardized electrical symbol list utilized in the electrical industry is provided by:
   A. the International Electrical Symbol Association.
   B. the Institute of Electrical and Electronics Engineers.
   C. the *National Electrical Code*.
   D. the Association of Electrical Designers.

3. Receptacles that are installed to serve wall outlets are those that will serve:
   A. specialty equipment.
   B. general purpose loads.
   C. appliance loads.
   D. special equipment loads.

4. Dedicated electrical circuitry is installed to serve:
   A. computers.
   B. photocopiers.
   C. vending machines.
   D. All of the above

5. The primary industry standard used in electrical design is:
   A. IESNA.
   B. EASA.
   C. *NEC*.
   D. NEMA.

You Are the Designer

Apply the knowledge you have gained from this previous chapter to your electrical design. In this section you will:
- Begin the first steps in developing an electrical design for a commercial building.
- Define the scope of your project.
- Develop a project folder for your design project.

■ About Your Project

To complete your task, you must know the following general details about your project:
- The scope of the plan, including the following:
  - Power plan with general and specialized equipment components
  - A lighting plan including lighting fixtures and controls
  - A distribution system including a main switchboard and panelboards
- Applicable standards from the *NEC* and National Electrical Manufacturers Association (NEMA)

■ Resources

To develop this part of the design, you need the following resources:
- The following documents from the *Student Resource CD-ROM:*
  - Plan sheet E_1: Interior building power
  - Plan sheet E_2: Building interior lighting
  - Plan sheet E_3: Single line diagram, developed panel schedules, load and short-circuit calculations
  - Electrical symbol list
- Three-ring binder with section dividers

■ Get to Work

At the onset of your project, you need to define the scope of your project. Begin by examining plan sheets E_1, E_2, and E_3 on the *Student Resource CD-ROM* and note all the required parts of the plan: general and specialized electrical requirements, lighting systems, and distribution system. Next, make a section for each component in a three-ring binder and add a section for general notes. Print out a copy of the electrical symbol list from the *Student Resource CD-ROM* as well and add this to the binder. The binder helps organize the documents you will gather, create, and reference throughout your design project. It also serves as a useful resource throughout the design process and helps you organize and compile the final design plan quickly and accurately.

CHAPTER 2

General Electrical Requirements

Chapter Outline

- Introduction
- Determining General Purpose Requirements
- Determining Branch-Circuit Requirements
- Designating Branch Circuits in the Electrical Plan

Learning Objectives

- Identify general purpose receptacle requirements and loads.
- Calculate general purpose receptacle branch-circuit loads.
- Create a panel schedule.
- Analyze, evaluate, and properly balance circuitry for general electrical loads.

Introduction

Almost every electrical design includes general electrical requirements, such as those for general purpose receptacle outlets (see **FIGURE 2-1**). Although these 120-volt receptacles (also known as convenience outlets) typically do not serve any specific type of equipment, they do serve a very important part in the overall design by providing power for general purpose devices and appliances.

Determining General Purpose Requirements

In commercial and industrial spaces, many 120-volt receptacles may be necessary to accommodate any devices that do not have special electrical requirements. The *National Electrical Code (NEC)* does not specify the quantity and location of these receptacles in commercial type occupancies, which ultimately depend on customer requirements.

If the end use customer is known during the design phase, the designer can customize the quantity and location of general purpose receptacles to meet the customer's specific needs. The designer might meet directly with the customer to discuss any specific requirements and alter the design accordingly. If the end use customer is not known, the designer must strive to create a design that provides for the average user's electrical needs and allows for flexibility for future additions and expansion. In this case, the design generally adheres to *NEC* requirements for general purpose receptacles in residential buildings [210.52(A)(1)] to provide a sufficient quantity of receptacles and meet any future needs. Though this section of the *Code* is not required for commercial and industrial buildings, the maximum advised distance between receptacles (12 linear feet) is a helpful guideline (see **FIGURE 2-2**).

Because general receptacle outlets are the least complex part of an electrical design, these receptacles are typically the first items designed. Referencing the floor plan, the designer adds general receptacles for general office spaces using the advised distance of 12 linear feet between receptacles. Additionally, the designer adds one receptacle for each restroom and one receptacle for approximately every 20 ft of wall space in hallways and corridors (to accommodate cleaning equipment such as vacuums). The *NEC* requires one 120-volt receptacle within 25 ft of any heating, air-conditioning, or refrigeration equipment, which may include any roof-mounted air-conditioning equipment [210.63]. For any other areas within the facility, the designer should consider how frequently the space is occupied and what the primary usage of each space is to determine the proper quantity and placement of general receptacle outlets.

Determining Branch-Circuit Requirements

After determining the general receptacle requirements, the next step in the design is to determine

FIGURE 2-1 Typical office spaces require one general purpose receptacle outlet every 10 to 12 ft.

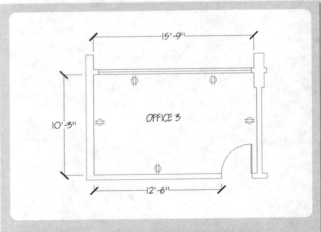

FIGURE 2-2 General purpose receptacles should be spaced approximately every 12 linear feet.

the proper number of required **branch circuits** to serve these receptacle loads. The type of circuit used for this purpose is a **general purpose branch circuit**, a circuit that serves two or more receptacles or outlets for lighting and appliances. Because these receptacles have no specific electrical requirements, the same branch circuit can serve several receptacles without **overload**.

As per the *NEC*, each receptacle is valued at 180 VA [220.14(I)]. In general, every ten 120-volt receptacles require one branch circuit at 80 percent of rated capacity. Although 120-volt general purpose receptacles are not considered **continuous loads**, and the branch circuits that serve them could be rated at 100 percent of their capacity (see Chapter 6), it is good practice to design branch circuits to serve these receptacles at 80 percent of their rated capacity as is advised for overcurrent protective devices, electrical panels, and related equipment that serves continuous loads. This reduces the risk that the branch circuits that serve these receptacles will become overloaded. If the designer does not follow this advice and designs each branch circuit at 100 percent of the load value, then each branch can serve approximately 13 receptacles. For most commercial applications, the 120-volt general purpose receptacle circuits are wired with conductors rated for 20 A and are served by single-pole 20-A overcurrent devices.

Please note that for multioutlet assemblies, such as those permanently fastened in place in plug strips, each 5 continuous linear feet of the assembly shall be rated at 180 VA for nonappliance use; when

---

Determining the Number of Allowable General Purpose 120-Volt Receptacles to Be Served by a 20-A Branch Circuit

To determine the number of allowable general purpose 120-volt receptacles to be served by a 20-A branch circuit:

1. Determine the allowable volt-amperes per circuit using Ohm's law.

    Ohm's law: Amperes (I) × Voltage (E) = Volt-amperes (VA)

    $$20 \text{ A} \times 120 \text{ V} = 2400 \text{ VA}$$

Answer: The circuit can serve 2400 VA.

2. Calculate the number of general purpose 120-volt receptacles that can be served by 2400 VA rated at full load value.

    $$\frac{\text{Allowable VA} \times \text{desired capacity (\%)}}{\text{VA rating}} = \text{Quantity of receptacles per circuit}$$

    $$\frac{2400 \text{ VA} \times 100\% \text{ [1.00]}}{180 \text{ VA}} = 13.3 \text{ receptacles}$$

Answer: At 100 percent of the branch circuit's rated volt-ampere capacity, it can serve approximately 13 receptacles.

3. Calculate the number of general purpose 120-volt receptacles that can be served by 2400 VA rated at 80 percent load value.

    $$\frac{\text{Allowable VA} \times \text{desired capacity (\%)}}{\text{VA rating}} = \text{Quantity of receptacles per circuit}$$

    $$\frac{2400 \text{ VA} \times 80\% \text{ [0.80]}}{180 \text{ VA}} = 10.6 \text{ receptacles}$$

Answer: At 80 percent of the branch circuit's rated volt-ampere capacity, it can serve approximately 11 receptacles.

utilized for appliance use, each 1 continuous linear foot shall be rated at a value of 180 VA [220.14(H)] (see **FIGURE 2-3**).

Balancing the Branch-Circuit Distribution

The branch-circuit layout must be designed for **balanced distribution**. Balanced distribution means that loads are designed with approximately equal volt-ampere values on each line or hot conductor that serves the panelboard serving the branch circuits. Balanced distribution allows for the best utilization of the equipment serving these loads. When loads are balanced equally among the lines (or phases) that serve them, the total load is distributed equally across the line wires (or phases), allowing the total load to be divided up with less load per line wire (or phase).

In commercial and industrial applications, balanced distribution reflects the daily usage for loads such as lighting that operate consistently during the day. It is more difficult to predict the usage of other loads such as devices operating on the general purpose 120-volt receptacles because it is highly unlikely that each 120-volt receptacle will be serving a load at one time. Even though the usage is inconsistent, these loads should be designed assuming a full-load condition with balanced distribution.

Most commercial and industrial office spaces are designed with a **120/208-volt, 3-phase, 4-wire wye system** that allows for the use of **multiwire branch circuits** to serve the general receptacles. When properly designed, this system provides for balanced distri-

TIP: The design standard is for all loads to be balanced to within 10 percent across each of the line wires (or phases) supplying the loads.

bution and ensures that should any one branch circuit fail, the space(s) served by these multiwire branch circuits would not be left without power to all the receptacles. This 3-phase, 4-wire system uses three ungrounded conductors and one grounded conductor (also called a neutral). Because in this system the voltage from each ungrounded phase conductor to the ground equals 120 volts, it is an ideal system for balanced distribution of general purpose loads. When the ungrounded branch circuit conductors are derived from different phases, the grounded conductor carries only the imbalance of the currents in the phase conductors. This allows for three individual branch circuits—one from each phase—to supply the general purpose loads with one common grounded conductor. Each 120-volt branch circuit should be designed to the same approximate volt-amperage and assigned circuit numbers derived from alternate phases, which balances the loads (see **FIGURE 2-4**).

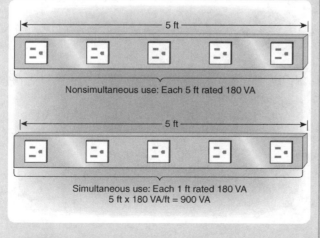

FIGURE 2-3 The requirements of 220.14(h)(l) and (H)(2) as applied to fixed multioutlet assemblies.

NFPA 70®, National Electrical Code® and NEC® are registered trademarks of the National Fire Protection Association, Quincy, MA.

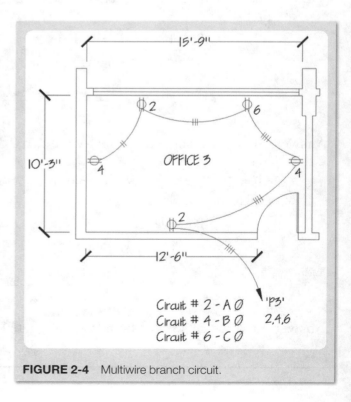

FIGURE 2-4 Multiwire branch circuit.

The receptacles in Figure 2-4 are served by a multiwire branch circuit. The circuit numbers 2, 4, and 6 are derived from alternate phases that allow for the balancing of the loads. When circuits are designed and wired using this method, each branch circuit can serve 10 receptacles for a total of 30 receptacles in each branch-circuit wiring run (10 for each circuit).

Panel Schedules

To assign and balance the circuits as they will be served from the distribution system, designers use a **panel schedule**, which is a representation of the actual panelboard that will serve the loads (see **FIGURE 2-5**). In panel schedules—as with panelboards—all the odd-numbered circuits are on the left and all the even-numbered circuits are on the right. When circuits are assigned horizontally across the panel schedule (e.g., circuits 1 and 2) they are assigned to the same phase. Therefore, if the designer wishes to alternate phases to balance the system, the circuits must be assigned vertically in the panel schedule (e.g., circuits 1, 3, and 5). In a 120/208-volt, 3-phase, 4-wire system, this means assigning the circuits to each of the A, B, and C phases. When multiwire branch circuits are assigned to 120/208-volt systems, they are done so in groups of three odd circuit numbers and then in groups of three even circuit numbers. This grouping method also allows the circuit breakers to be located vertically adjacent to each other in the panelboard.

In addition to the branch-circuit numbers, panel schedules can also provide information about the product itself, including:
- Panel size
- Panel voltage
- Quantity of the branch-circuit overcurrent devices
- Quantity of the devices served by the branch circuits
- Panel identification
- Location

PANEL	P3'															
LOCATION:	Equipment Room						VOLTAGE / PHASE:		120/208 3PH. 4 W.							
FLOOR:	First						BUSS:		225A							
MOUNTING:	Surface						MAIN BREAKER:		200A							

	CT. #	OUTLETS			VOLT AMPS			BKR/ POLE	BKR/ POLE	VOLT AMPS			OUTLETS			CT. #	
		LTG	REC	MISC	A	B	C			A	B	C	MISC	REC	LTG		
Receptacles	1.		10		1980			20/1	20/1	1800				10		2.	Receptacles
	3.		10			1800					1620			9		4.	
	5.		10				1800					1620		9		6.	
	7.		9		1620					1800			1			8.	Vend. Machine
	9.		9			1620					1800		1			10.	
	11.		9				1620					1800	1			12.	
Microwave L.R.	13.			1	1200					1260				7		14.	Receptacles
Copy Mach. 1	15.			1		1440					1080			6		16.	
Copy Mach. 2	17.			1			1440					1080		6		18.	
Receptacles	19.		9		1620					720				4		20.	Receptacle
	21.		9			1620					720			4		22.	
	23.		8				1440					900		5		24.	
Receptacles Roof	25.		2		1920					1920				6		26.	Receptacle / Module
	27.		2			1920					1920			5		28.	
	29.		2				1920					1920		5		30.	
Receptacle/Module	31.		8		1920											32.	Spare
	33.		8			1920										34.	
	35.		8				1920									36.	
Network Server	37.				1200											38.	
	39.					1200										40.	
	41.						1200									42.	
				sub total	11460	11520	11340			7500	7140	7320	sub total				

Total A Phase	=	18960 va			
Total B Phase	=	18660 va			
Total C Phase	=	18660 va			
LCL (25%)	=	va			
Total load		56280 va /	360	156 Amps	

FIGURE 2-5 A 120/208-volt, 3-phase, 4-wire panel schedule.

- Type of mounting
- Voltage
- Phase
- Capacity in amperes
- Type of load served
- Load of the branch circuits, each phase, and the total load in both volt-amperes and amperage
- Number of poles of the branch-circuit overcurrent device serving each of the individual branch circuits

Electrical contractors can submit copies of panel schedules to suppliers for pricing estimates to potentially eliminate the need for them to compile separate material lists of panelboards and their components.

Designating Branch Circuits in the Electrical Plan

To properly reference which branch circuit serves each receptacle on the design plan, the specified branch circuit for each receptacle must be shown on the print at each receptacle location. The designer may also indicate in the design plan the exact routing of the **raceways** that serve these loads. If the raceways are not shown, then the routing and installation of the raceways are left to the discretion of the installer (see **FIGURE 2-6**). When designers use a design method where they illustrate both the branch-circuit numbers and the raceway design on the plan, installers must not deviate from the assigned circuits and raceway routing in the field.

Should the installer need to deviate from the design plan because of variations and/or changes in the field, a set of "as-built" drawings should be submitted that indicate any changes that were made to the original plan.

When the raceways are indicated in the design, designers identify the number of conductors in each raceway using slash marks (see **FIGURE 2-7**). This type of design includes how the raceways are installed for each receptacle and reference which raceway, known as the **home run**, will be installed back to the source. The home run raceway is identified with an arrow symbol and includes notation indicating which branch circuits will be installed in the raceway and which branch-circuit panelboard

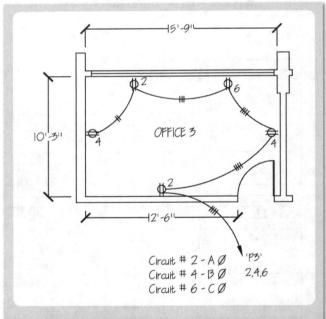

FIGURE 2-7 Branch-circuit identification including raceways, conductors, and home run detail.

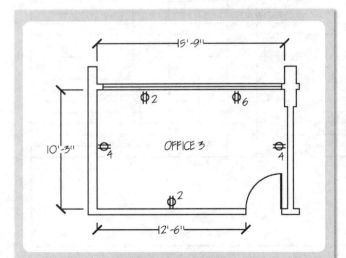

FIGURE 2-6 Branch-circuit identification for general purpose receptacles.

Plans that show the raceway layout can be helpful to electrical technicians who need to make changes or repairs after the installation.

will serve these branch circuits. In the example in Figure 2-4, the home run indicates that four conductors assigned to circuit numbers 2, 4, and 6 are to be run back to panelboard P_3.

Unless the designer specifies the raceway and grounding methods, the installer can choose any raceway method approved by the *NEC*. Not all raceway methods are approved for use as a grounding method; therefore, an additional grounding conductor must be installed. Because the raceway method used may vary, the illustration of the number of conductors to be installed in a particular raceway typically does not include any references to an additional grounding conductor that may be necessary. The installer makes that determination for the raceway type based on knowledge of wiring practices and the requirements of the *NEC*.

As mentioned, depending on the raceway method used, installation of an additional equipment grounding conductor may be required. In this case, an additional equipment grounding conductor would not be required and is therefore not shown in the design. The equipment grounding conductor should be illustrated only when the raceway type chosen requires the additional conductor.

When designing with multiwire circuits that will share a common ungrounded conductor, it is good design practice to locate the circuit breakers side by side in a panelboard in a vertical sequence to create balanced distribution. In the 3-phase panelboard, this ensures that each branch circuit is assigned to an alternate phase; in the case of a single-phase distribution system, this method ensures that each branch circuit is assigned to alternate line wires. According to the *NEC*, at the point where a multiwire branch circuit receives its supply, it must have a means to simultaneously disconnect all ungrounded conductors [210.4(B)].

When more than one branch circuit serves two devices on a single mounting strap (yoke), as in the case with some split duplex receptacles, the *NEC* requires that circuit breakers serving the device must have their handles tied together [210.7(B)]. This method avoids the possibility of electrical shock during servicing by ensuring that if one branch circuit is disconnected, any other branch circuits serving the same device are also disconnected. Alternating the branch circuits in a panelboard for balanced distribution and aligning the circuit breakers in a vertical position enable multiwire branch circuits to conform to 210.(4)(B) and 210.7(B) and the handles of the circuit breakers to be tied together by an approved method. If the circuit breakers for the multiwire branch circuits were located across from each other in the panelboard, they cannot physically be tied together. Designers must consider conforming to these requirements when designing home runs.

Wrap Up

■ Master Concepts

- All commercial and industrial offices have basic electrical equipment requirements that can be served with the general purpose 120-volt branch circuits through general purpose wall-mounted receptacle outlets.
- General purpose electrical loads have no special requirements; therefore, several receptacle outlets may be served by the same branch circuit.
- All electrical loads in commercial or industrial buildings must be designed for balanced distribution.
- Panel schedules (representations of the actual panelboard that provides distribution of the electrical supply) are used during the design phase to simulate how the loads will be distributed. A panel schedule is an excellent tool for achieving design goals to actual field conditions.
- Each branch circuit serving general receptacle outlets or equipment is identified on an electrical plan at the receptacle or equipment location by its branch-circuit number. This information helps installers and is useful for servicing.
- Proper identification of the circuit allows installers to properly wire the device to the correct circuit and service technicians to properly identify and de-energize the circuit, providing a greater degree of safety.

■ Charged Terms

Many of the definitions supplied here are from the *NEC*.

120/208-volt, 3-phase, 4-wire, wye system A distribution system generated with three individual sine waves separated by 120 electrical degrees that are identified as phases A, B, and C. One leg of each of the three phase coils is electrically connected to the others at a common point, forming a wye, which when grounded, becomes the fourth wire (or neutral) in the system. This allows for each of the three individual phase voltages to supply 120 volts to the grounded point, while the line voltage across each of the phases produces 208 volts. The line-to-line voltages can supply both 208-volt 3-phase and 208-volt single-phase. Because the three individual phases each can supply 120 volts, this system is commonly used in commercial office applications where 120 volts is desired because the 120-volt loads can be balanced across each of the three phases.

balanced distribution An electrical distribution system in which the ungrounded conductors carry equal currents. In distribution systems that also include a grounded conductor, the grounded conductor will carry the imbalance of the currents in the ungrounded conductors.

branch circuit The circuit conductors between the final overcurrent device protecting the circuit and the outlet(s) [100].

continuous load A load where the maximum current is expected to continue for three hours or more [100].

general purpose branch circuit A branch circuit that supplies two or more receptacles or outlets for lighting and appliances [100].

home run The raceway designated on a plan as the one that carries branch-circuit conductors back to the serving source (such as a panelboard).

multiwire branch circuit A branch circuit that consists of two or more ungrounded conductors that have a voltage between them, and a grounded conductor that has equal voltage between it and each ungrounded conductor of the circuit and that is connected to the neutral or grounded conductor of the system [100].

overload Operation of equipment in excess of normal, full-load rating, or of a conductor in excess of rated ampacity that, if it persists for a sufficient length of time, would cause damage or dangerous overheating. A fault, such as a short circuit or ground fault, is not an overload [100].

panel schedule An illustration of key panelboard information showing how branch circuitry is distributed, number of phases, voltage, and size in amperage; panel schedules are completed by hand calculation or by computer software.

raceway An enclosed channel of metal or nonmetallic materials designed expressly for holding wires, cables, or busbars [100].

receptacle outlet An outlet where one or more receptacles are installed [100].

■ Check Your Knowledge

1. In the basic layout for general purpose receptacles, the receptacles should be spaced at a linear distance of approximately _____ from each receptacle.
   - A. 6 ft
   - B. 12 ft
   - C. 4 ft
   - D. 8 ft

2. As per the *National Electrical Code,* each 120-volt general purpose receptacle installed in a commercial application must be rated at:
   - A. 360 VA.
   - B. 120 VA.
   - C. 180 VA.
   - D. 540 VA.

3. A 120-volt 20-A branch circuit can serve _____ general purpose receptacles.
   - A. 2
   - B. 10 to 13
   - C. 15 to 17
   - D. up to 20

4. A multioutlet assembly serving general loads has a load value based on _____ VA per five linear feet.
   A. 120
   B. 180
   C. 360
   D. 540

5. How many conductors are required for a raceway serving a general purpose, 120-volt multiwire branch circuit with three branch circuits served from a 120/208-volt 3-phase 4-wire system, assuming that the raceway is an approved equipment grounding conductor?
   A. 3
   B. 4
   C. 5
   D. 6

6. A 120-volt general purpose receptacle must have the branch circuit that serves the receptacle identified at what point on the design plan?
   A. At the panelboard that serves the branch circuit
   B. At the receptacle location, by circuit number
   C. It is left to the discretion of the designer.
   D. No identification is necessary.

7. When referencing an electrical plan, a home run is the:
   A. branch-circuit number that serves the devices.
   B. raceway installed from the serving panelboard to the first device.
   C. raceway connected to the last device in the run.
   D. conduit that contains the feeders that supply a panelboard.

8. General purpose 120-volt receptacles provided for roof-mounted equipment must be located within _____ of the equipment.
   A. 5 ft
   B. 10 ft
   C. 25 ft
   D. 100 ft

9. As-built drawings are utilized to:
   A. differentiate between the original design plan and the completed structure.
   B. illustrate how the building is to be constructed.
   C. provide further details about a project.
   D. provide a reference when designing plans for a project requiring alterations.

10. When more than one branch circuit serves two devices on a single mounting strap:
    A. the circuit breakers serving the devices must be identified at the panelboard.
    B. the branch circuits that supply the devices must be identified at the location of the devices.
    C. the circuit breakers serving the devices must be provided with a method to physically join the circuit breaker handles together.
    D. Both A and B

You Are the Designer

Apply the knowledge you have gained from this previous chapter to your own electrical design. In this section you will:
- Design 120-volt general purpose receptacles
- Design branch circuits serving the receptacles
- Illustrate the following on the design plan:
  - General purpose receptacle outlets
  - Raceways serving each receptacle
  - Quantity of conductors contained in each raceway
  - Branch circuits serving each receptacle
  - Home run raceway
- Develop a panel schedule to assign branch circuits for the 120-volt general purpose receptacles

■ About Your Project

To complete your task, you must know the following details about your project:
- Which areas within the facility require general purpose receptacles (e.g., general office areas, entry and reception areas, corridors, restrooms)

■ Resources

To develop your part of the design, you need the following resources:
- The blank plan sheet E_1 found in the text.
- The following documents from the *Student Resource CD-ROM*:
  - Completed panel schedule P_3
  - Blank panel schedule template
  - Completed plan sheet E_1

■ Get to Work

Installing one receptacle outlet for every 10 to 12 feet of linear wall space provides for an adequate number of general purpose 120-V receptacles.

General Receptacles

You first need to determine all the general purpose 120-volt receptacle outlets necessary for your project. Before you begin, review the completed plan sheet E_1 and locate all the areas on the completed plan where these receptacles were designed. Start your design by drawing the appropriate symbol in all of the locations you determine require receptacles. Remember that although there are no specific spacing requirements between receptacles, a good design practice is to provide one receptacle for approximately every 10 to 12 linear feet of wall space. No specific mounting height is listed, but standard practice is, unless noted otherwise, to mount receptacles 12 in. from finished floor to the receptacle centerline. When receptacles are to be installed at a different height (as in restrooms and above counter spaces), the plan should include notation next to the receptacle (such as +42″, which indicates that the receptacle should be installed 42 in. from finished floor to receptacle centerline) (See **FIGURE 2-8**).

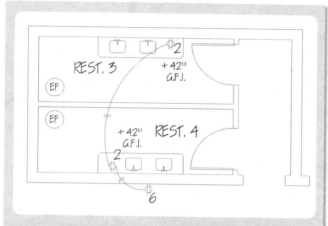

FIGURE 2-8 For general purpose receptacles that will be mounted at heights other than the standard 12 in. to centerline, a reference to the mounting height must be noted on the plans.

All receptacles installed in restrooms or outdoors must be GFCI (ground-fault circuit-interrupter) protected [210.8(B)].

For this project, the receptacles for the reception area counter spaces have a closer distance between receptacles to serve the needs of the counter area for general electrical equipment such as computers. These receptacles were mounted at the typical distance of 12 in. from floor to center of the outlets, not above the counter as done in many older designs. In many modern commercial environments, plug strips are necessary; when located below the counter, a hole with a finished trim piece is cut in the counter to provide access to the receptacles below.

Note that on the completed plan, a symbol indicates that receptacles are to be floor mounted in the showroom area. Also notice the designation of the underground raceways to serve them. These floor-mounted receptacles were designed to accommodate display cases installed for product display located away from wall-mounted receptacles. Display cases often require receptacles for lighting fixtures that are components of the case; the dotted line reference indicates that raceways serving the floor-mounted receptacles are to be served with an underground raceway. The underground raceway terminates at a wall-mounted receptacle, which provides access to the raceway.

All the exterior walls for the showroom area, with the exception of one, are floor-to-ceiling glass wall panels that do not allow wall-mounted receptacles to be installed. This is also the case for the glass wall panel in the reception area; for this area, a floor-mounted receptacle is provided.

In the lunch room area, the receptacles illustrated with the branch-circuit designations 8, 10, 12, and 13, are required for vending machines and a counter-mounted microwave unit. Do not install any receptacle outlets for these devices at this time; these are specialty devices that require more specialized circuitry and you will design them in Chapter 3.

After you complete the design for the general receptacle requirements for the office areas, your next task is to design the general purpose receptacles for the manufacturing area. These receptacles are all typically installed as the office area receptacles are, for general purposes with no special requirements. Machinery operators may have small hand-powered tools that require 120 volts; therefore, it is advisable to provide a receptacle in the general area of each machine. The assembly area has several work benches that should have several additional receptacles installed at 42 in. above the finished floor, as noted on the completed plan.

Before moving on to the next step, review the plan carefully for any additional receptacles you may want to add or any areas that may have been overlooked. Add any additional receptacles now before branch circuits are assigned and a preliminary panel schedule is developed because additional receptacles affect the branch-circuit number layout and the calculated loads for each of the branch circuits.

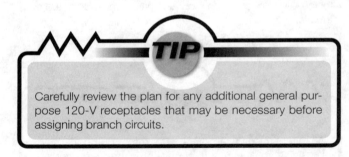

Carefully review the plan for any additional general purpose 120-V receptacles that may be necessary before assigning branch circuits.

Branch Circuits

The next step of the design is to determine which branch circuits will serve all the general purpose receptacles in your design plan. From this determination, you then identify each branch circuit that is to serve each of the receptacles at its respective location on the plan.

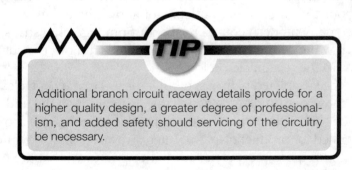

Additional branch circuit raceway details provide for a higher quality design, a greater degree of professionalism, and added safety should servicing of the circuitry be necessary.

For this design, the office portion of the project is served by a typical 120/208-volt, 3-phase, 4-wire system and 120-volt multiwire branch circuits. Following the Calculation Examples presented earlier in this chapter, locate areas in the office space where approximately 30 to 39 receptacle outlets (that is, 10 to 13 per branch circuit) can be served by the same raceway in the same approximate area. Once you complete this for the entire plan, go back and assign branch-circuit numbers to each of the receptacles in alternating groups of three odd or even numbers (e.g., 1, 3, 5 or 2, 4, 6), allowing for approximately 10 to 13 receptacles to be served by each branch circuit (see **FIGURE 2-9**).

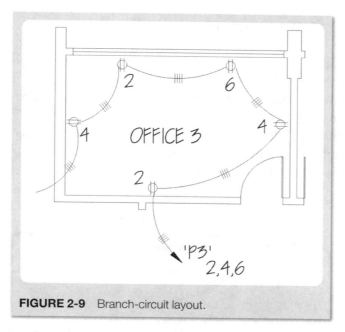

FIGURE 2-9 Branch-circuit layout.

To see an example, reference the completed plan and completed panel schedule P_3. These alternating branch-circuit numbers were used in the office areas and allowed for the balancing of the loads across the three phases. Once your branch circuits have served the maximum number allowed, continue with the next group of three alternating branch-circuit numbers (e.g., 7, 9, 11). Continue the design until all the receptacles have been assigned an appropriate branch-circuit number.

Next, route the raceways and designate the home runs. Although it is not a requirement that all the raceway and home run details be illustrated on the plan, providing all the additional detail creates a higher quality design and demonstrates a higher degree of professionalism. If you choose not to include all the detail for the raceways, you must at least illustrate the appropriate branch-circuit number at each receptacle location. The more detail that you supply helps those who may be providing a proposal for the project and installers in the field.

The next step is to determine which receptacle for each of the branch circuits will be the home run. Locate each raceway run supplying each of the three groups of branch-circuit numbers closest to the serving panelboard and designate the home run at this point. Use the completed plan as a reference. On the completed plan, home run raceways were designed as close as practically possible to the serving panelboard (in Office 6, the equipment room). Although not a requirement, this design method can eliminate unnecessary raceway materials to help save time and costs.

Panel Schedules

The final step in the branch-circuit design is to enter the branch circuits into your panel schedule. Using the template on the *Student Resource CD-ROM*, identify a title for each of the branch circuits. Use the completed panel schedule P_3 as a reference; notice how circuits 1, 2, 3, and others are labeled "Receptacles." For each of the branch circuits, count the number of assigned receptacles and multiply by 180 VA to obtain the total volt-ampere load for each branch circuit.

For example, in panel schedule P_3 on the completed plan, notice that for branch-circuit numbers 1, 3, and 5, each branch circuit serves 10 receptacles. Multiply this by 180 VA for a total of 1800 VA. Branch circuits 7, 9, and 11 each serve 9 receptacles. Multiply this by 180 VA to determine the total volt-ampere rating of 1620 VA.

Complete the panel schedule by entering all the branch-circuit numbers you have assigned. Enter these numbers into the panel schedule under the heading of "OUTLETS/REC" along with the number of receptacle outlets served by each branch circuit. When you enter these data into the template, the total for each of the three A, B, and C phases is automatically calculated at the bottom of the panel schedule. This helps ensure that the phase is balanced; if the end phase totals are out of balance, you should reassign branch circuits on the plan and in the panel schedule until your design indicates the best balancing across the phases.

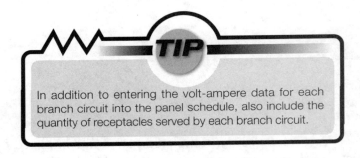

TIP

In addition to entering the volt-ampere data for each branch circuit into the panel schedule, also include the quantity of receptacles served by each branch circuit.

To complete this portion of the panel schedule, enter the proper size overcurrent protective device (circuit breaker) that will serve each branch circuit. As mentioned, the typical circuit breaker size for 120-V general purpose branch circuits is a 20 ampere single-pole device. Enter this value into the panel schedule under the heading of BRKR/POLE as 20/1 (see Completed Panel Schedule P_3).

CHAPTER 3

Specialized Electrical Requirements

Chapter Outline

- **Introduction**
- **Equipment List**
- **Specialized Equipment Branch Circuits**
- **Motors**
- **Raceways for Branch Circuit-Distribution**
- **Determining the Number of Panelboards**
- **Raceway Legends**

Learning Objectives

- Create an equipment list for a set of electrical plans.
- Determine the required number of branch circuits needed for specialized equipment.
- Apply relevant *National Electrical Code (NEC)* requirements to motor applications, including their equipment branch-circuit conductors, short-circuit and ground-fault protection devices, and grounding techniques.
- Calculate and design required branch circuits and raceway sizes for equipment in a design plan.
- Create a panel schedule and raceway legend.

Introduction

Specialized equipment loads (including manufacturing and specialized office equipment such as photocopiers) have more design requirements than do general purpose loads, including design of individual branch circuitry and specialized grounding methods that may be required. Therefore, greater emphasis is placed on the specific details, required calculations, and *NEC* regulations for these types of loads than for general electrical loads.

Equipment List

The owner or architect should give the electrical designer **manufacturer electrical specification sheets** for all the specialized equipment in a given facility along with a floor plan showing the location of all equipment in the facility so that the electrical designer can create an **equipment list**. The equipment list determines the electrical requirements of specialized equipment and machinery (see **FIGURE 3-1**). To create this list, the designer must gather the following information:

1. A brief description of each piece of equipment
2. Specific electrical requirements for each piece of equipment, including:
   a. Horsepower
   b. Voltage
   c. Number of phases
   d. Load (in amperes or kVA)
3. Floor plan indicating the location of each piece of equipment

Once the designer compiles the equipment list, he or she should organize the equipment by type so that each piece can be easily referenced with its location on the floor plan. The designer then calculates the required number of branch circuits necessary to serve each piece of equipment. Later in the design process, the designer will use these totals to determine the quantity and operating voltages of panelboards to serve the equipment loads.

Specialized Equipment Branch Circuits

As per the *NEC*, specialized equipment must be served by separate branch circuits to avoid overload [430]. The most basic type of separate branch circuit is a general branch circuit that can serve small office equipment such as photocopiers, computer equipment, and vending machines.

There are three common types of specialized equipment branch circuit wiring methods. In the first type, the equipment is served by its own ungrounded conductor and the grounded conductor is shared with other circuits such as a multiwire branch circuit (see **FIGURE 3-2**). In the second type, the grounded conductor will not be shared with any other branch circuits; this type of branch circuit provides better electrical isolation and is therefore often used for electrically sensitive equipment (see **FIGURE 3-3**). In the third type of separate circuit, all the branch-circuit conductors supplying an individual piece of equipment will not share any conductors with any other equipment or receptacles and, in some cases, will not share a raceway with any other circuit conductors. This type of circuit contains its own ungrounded, grounded, and, if applicable, **equipment grounding conductors** (see **FIGURE 3-4**).

It is important to note that the *NEC* does not recognize flexible metallic conduit as an approved grounding method when installed lengths exceed 6 feet and are protected by overcurrent devices greater than 20 A [250.118(5)]; in such cases, a supplemental equipment grounding conductor is

If the design is being completed for a preexisting building that is undergoing alterations and manufacturer electrical specification sheets are not available, the designer may need to visit the location of the existing equipment to obtain the electrical requirements for each piece.

Electrical requirements for specialized equipment will be specified by the equipment manufacturer and should be noted on the plans in an equipment list.

EQUIPMENT LIST

#	Description	Ratings	#	Description	Ratings
1	CNC Lathe 1	208V 3Ø 5 H.P. 16.7 F.L.A.	20	A/C-2	480V 3Ø 15 H.P. 21.0 F.L.A.
2	CNC Lathe 2	208V 3Ø 7.5 H.P. 24.2 F.L.A.	21	A/C-3	480V 3Ø 10 H.P. 14.0 F.L.A.
3	CNC Lathe 3	208V 3Ø 7.5 H.P. 24.2 F.L.A.	22	A/C-4	480V 3Ø 10 H.P. 14.0 F.L.A.
4	CNC Lathe 4	208V 3Ø 5 H.P. 16.7 F.L.A.	23	A/C-5	480V 3Ø 15 H.P. 21.0 F.L.A.
5	CNC Lathe 5	480V 3Ø 15 H.P. 21.0 F.L.A.	24	A/C-6	480V 3Ø 15 H.P. 21.0 F.L.A.
6	CNC Lathe 6	480V 3Ø 15 H.P. 21.0 F.L.A.	25	Exaust Fan Roof	208V 3Ø 1 H.P. 4.6 F.L.A.
7	CNC Lathe 7	480V 3Ø 15 H.P. 21.0 F.L.A.	26	Vend. Mach.	120V 1Ø 1800VA 15.0 F.L.A.
8	CNC Lathe 8	480V 3Ø 15 H.P. 21.0 F.L.A.	27	Vend. Mach.	120V 1Ø 1800VA 15.0 F.L.A.
9	Grinder 1	208V 1Ø 3 H.P. 18.7 F.L.A.	28	Vend. Mach.	120V 1Ø 1800VA 15.0 F.L.A.
10	Grinder 2	208V 1Ø 1.5 H.P. 11.0 F.L.A.	29	Microwave	120V 1Ø 1200VA 10.0 F.L.A.
11	Grinder 3	208V 1Ø 3 H.P. 18.7 F.L.A.	30	Copier 1	120V 1Ø 1440VA 12.0 F.L.A.
12	Grinder 4	208V 1Ø 1.5 H.P. 11.0 F.L.A.	31	Copier 2	120V 1Ø 1440VA 12.0 F.L.A.
13	Grinder 5	208V 3Ø 3 H.P. 18.7 F.L.A.	32	Fork Lift Charger	120V 1Ø 1200VA 10.0 F.L.A.
14	Extruder	480V 3Ø 50 H.P. 65.0 F.L.A.	33	Fork Lift Charger	120V 1Ø 1200VA 10.0 F.L.A.
15	CNC Mill 1	480V 3Ø 15 H.P. 21.0 F.L.A.	34	Mod. Furn. Panels	120V 1Ø 1200VA 10.0 F.L.A.
16	CNC Mill 2	480V 3Ø 25 H.P. 34.0 F.L.A.	35	Network Server	120V 1Ø 1200VA 10.0 F.L.A.
17	CNC Mill 3	480V 3Ø 15 H.P. 21.0 F.L.A.	36	Network Server	120V 1Ø 1200VA 10.0 F.L.A.
18	CNC Lathe 9	480V 3Ø 10 H.P. 14.0 F.L.A.	37	Network Server	120V 1Ø 1200VA 10.0 F.L.A.
19	A/C-1	480V 3Ø 15 H.P. 21.0 F.L.A.			

FIGURE 3-1 The equipment list provides details about all specialized equipment and their electrical requirements.

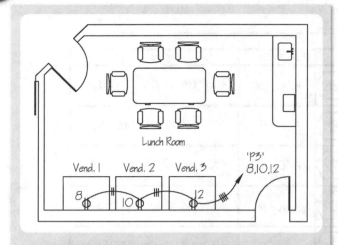

FIGURE 3-2 Individual branch circuits that share a common grounded conductor.

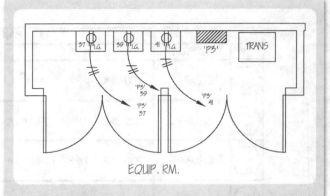

FIGURE 3-4 Individual branch circuits that are supplied with their own ungrounded, grounded, separate grounding conductor, and individual raceways.

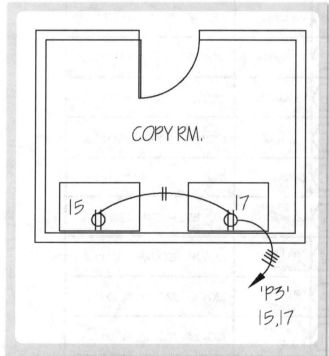

FIGURE 3-3 Individual branch circuits that do not share a common grounded conductor.

needed. For the third type of specialty circuit, the equipment must have its own grounding conductor; therefore, the raceway system will contain two equipment grounding conductors—one for the equipment and one for grounding the raceway. If the ground terminal of a receptacle is to serve equipment only and is separate from a grounding conductor utilized to ground a raceway, it must be listed and labeled as an approved isolated ground type. Designers identify these receptacles on a print as <u>isolated ground (IG)</u>.

■ Motors

Motors of 1 hp and greater (typically found in production machinery) must be provided with specialized branch circuits. As per the *NEC*, designers must determine the sizing of branch-circuit conductors, sizing of the devices protecting the motors, and wiring serving these devices [Article 430]. Designers can make these determinations using *NEC* guidelines, the equipment list, and the following steps:

1. Determine the size of the motor equipment branch-circuit conductors based on the motor voltage and full-load amperage.
2. Determine the size of the overcurrent protective devices for the motor and associated wiring.
3. Determine equipment grounding methods.
4. Select and size raceway method(s).

With all types of specialized equipment branch circuits, any special equipment grounding requirements must be noted on the plans either at the device being served or in a keynotes section.

Motor Equipment Branch-Circuit Conductors

In most applications, control devices start motors by applying full voltage to the terminals of the motor when the devices are energized. A motor's starting current can be up to six times its running current. The NEC requires that the size of the branch-circuit conductors serving a single motor in a continuous duty application be based on a minimum size of 125 percent of the motor's full-load current [430.22(A)]. By basing the conductor size on this calculated amperage designers size conductors larger than is needed for full-load amperage, which helps handle higher **inrush currents** associated with motors.

If the branch-circuit conductors serving the equipment will be more than 100 ft from the electrical source to the equipment, **voltage drop** (loss resulting from the length of the conductor, its resistance, and the amperage imposed on the conductor) may become a factor. Low voltage at the equipment can damage or greatly shorten the life of a motor. Designers must calculate voltage drop when a motor's branch-circuit conductors are 100 ft or greater in length. These calculations ensure that conductors have adequate capacity to serve both the voltage and amperage necessary for the equipment to operate correctly and efficiently.

According to the NEC:
Conductors for branch circuits . . . [should be] sized to prevent a voltage drop exceeding 3 percent at the farthest outlet of power, heating, and lighting loads, or combinations of such loads, and where the maximum total voltage drop on both feeders and branch circuits to the farthest outlet does not exceed 5 percent, provide reasonable efficiency of operation [210.19(1) FPN No. 4].

TABLE 3-1 outlines the maximum allowable voltage drop for branch circuits based on the NEC guidelines.

TABLE 3-1 Maximum Allowable Voltage Drop for Branch Circuits

Branch-Circuit Voltage	3% Voltage Drop	5% Voltage Drop
120 volts	3.6 volts	6 volts
208 volts	6.24 volts	10.4 volts
277 volts	9.23 volts	13.85 volts
480 volts	14.4 volts	20.4 volts

The following NEC tables are valuable references for sizing motor equipment branch-circuit conductors:
- Table 310.16 Allowable Ampacities of Insulated Conductors
- Table 430.247 Full Load Currents in Amperes, Direct Current Motors
- Table 430.248 Full Load Currents in Amperes, Single Phase Alternating Current Motors
- Table 430.250 Full Load Current, Three Phase Alternating Current Motors

To calculate the correct wire size to account for voltage drop in branch circuits, use the following equations.

For single-phase systems:
$$Cm = \frac{K \times D \times I \times 2}{Vd}$$

For 3-phase systems:
$$Cm = \frac{K \times D \times I \times 1.73}{Vd}$$

where
Cm = Circular mil area of conductor
K = DC wire constant
D = One-way distance
I = Load in amperes
Vd = Allowable voltage drop

The DC wire constant value (K) is a measure of a conductor's DC ohmic value resistance for a conductor size of exactly 1 mil wide by 1 ft long. For calculating voltage drop, the DC wire constant value (K) is 12.9 for copper conductors and 21.2 for aluminum conductors. To determine the circular mil area of the conductor (Cm), refer to **TABLE 3-2**, which is NEC Table 8.

Voltage drop calculations should be performed whenever branch-circuit conductors will serve a load located 100 ft or more from the source.

TABLE 3-2 *NEC* Table 8 Conductor Properties

Size (AWG or kcmil)	Area		Conductors						
			Stranding			Overall			
				Diameter		Diameter		Area	
	mm²	Circular mils	Quantity	mm	in.	mm	in.	mm²	in.²
18	0.823	1620	1	—	—	1.02	0.040	0.823	0.001
18	0.823	1620	7	0.39	0.015	1.16	0.046	1.06	0.002
16	1.31	2580	1	—	—	1.29	0.051	1.31	0.002
16	1.31	2580	7	0.49	0.019	1.46	0.058	1.68	0.003
14	2.08	4110	1	—	—	1.63	0.064	2.08	0.003
14	2.08	4110	7	0.62	0.024	1.85	0.073	2.68	0.004
12	3.31	6530	1	—	—	2.05	0.081	3.31	0.005
12	3.31	6530	7	0.78	0.030	2.32	0.092	4.25	0.006
10	5.261	10380	1	—	—	2.588	0.102	5.26	0.008
10	5.261	10380	7	0.98	0.038	2.95	0.116	6.76	0.011
8	8.367	16510	1	—	—	3.264	0.128	8.37	0.013
8	8.367	16510	7	1.23	0.049	3.71	0.146	10.76	0.017
6	13.30	26240	7	1.56	0.061	4.67	0.184	17.09	0.027
4	21.15	41740	7	1.96	0.077	5.89	0.232	27.19	0.042
3	26.67	52620	7	2.20	0.087	6.60	0.260	34.28	0.053
2	33.62	66360	7	2.47	0.097	7.42	0.292	43.23	0.067
1	42.41	83690	19	1.69	0.066	8.43	0.332	55.80	0.087
1/0	53.49	105600	19	1.89	0.074	9.45	0.372	70.41	0.109
2/0	67.43	133100	19	2.13	0.084	10.62	0.418	88.74	0.137
3/0	85.01	167800	19	2.39	0.094	11.94	0.470	111.9	0.173
4/0	107.2	211600	19	2.68	0.106	13.41	0.528	141.1	0.219
250	127	—	37	2.09	0.082	14.61	0.575	168	0.260
300	152	—	37	2.29	0.090	16.00	0.630	201	0.312
350	177	—	37	2.47	0.097	17.30	0.681	235	0.364
400	203	—	37	2.64	0.104	18.49	0.728	268	0.416
500	253	—	37	2.95	0.116	20.65	0.813	336	0.519
600	304	—	61	2.52	0.099	22.68	0.893	404	0.626
700	355	—	61	2.72	0.107	24.49	0.964	471	0.730
750	380	—	61	2.82	0.111	25.35	0.998	505	0.782
800	405	—	61	2.91	0.114	26.16	1.030	538	0.834
900	456	—	61	3.09	0.122	27.79	1.094	606	0.940
1000	507	—	61	3.25	0.128	29.26	1.152	673	1.042
1250	633	—	91	2.98	0.117	32.74	1.289	842	1.305
1500	760	—	91	3.26	0.128	35.86	1.412	1011	1.566
1750	887	—	127	2.98	0.117	38.76	1.526	1180	1.829
2000	1013	—	127	3.19	0.126	41.45	1.632	1349	2.092

Notes:
1. These resistance values are valid **only** for the parameters as given. Using conductors having coated strands, different stranding type, and, especially, other temperatures changes the resistance.
2. Formula for temperature change: $R_2 = R_1 [1 + \alpha (T_2 - 75)]$ where $\alpha_{cu} = 0.00323$, $\alpha_{AL} = 0.00330$ at 75°C.

Direct-Current Resistance at 75°C (167°F)					
Copper				Aluminum	
Uncoated		Coated			
ohm/km	ohm/kFT	ohm/km	ohm/kFT	ohm/km	ohm/kFT
25.5	7.77	26.5	8.08	42.0	12.8
26.1	7.95	27.7	8.45	42.8	13.1
16.0	4.89	16.7	5.08	26.4	8.05
16.4	4.99	17.3	5.29	26.9	8.21
10.1	3.07	10.4	3.19	16.6	5.06
10.3	3.14	10.7	3.26	16.9	5.17
6.34	1.93	6.57	2.01	10.45	3.18
6.50	1.98	6.73	2.05	10.69	3.25
3.984	1.21	4.148	1.26	6.561	2.00
4.070	1.24	4.226	1.29	6.679	2.04
2.506	0.764	2.579	0.786	4.125	1.26
2.551	0.778	2.653	0.809	4.204	1.28
1.608	0.491	1.671	0.510	2.652	0.808
1.010	0.308	1.053	0.321	1.666	0.508
0.802	0.245	0.833	0.254	1.320	0.403
0.634	0.194	0.661	0.201	1.045	0.319
0.505	0.154	0.524	0.160	0.829	0.253
0.399	0.122	0.415	0.127	0.660	0.201
0.3170	0.0967	0.329	0.101	0.523	0.159
0.2512	0.0766	0.2610	0.0797	0.413	0.126
0.1996	0.0608	0.2050	0.0626	0.328	0.100
0.1687	0.0515	0.1753	0.0535	0.2778	0.0847
0.1409	0.0429	0.1463	0.0446	0.2318	0.0707
0.1205	0.0367	0.1252	0.0382	0.1984	0.0605
0.1053	0.0321	0.1084	0.0331	0.1737	0.0529
0.0845	0.0258	0.0869	0.0265	0.1391	0.0424
0.0704	0.0214	0.0732	0.0223	0.1159	0.0353
0.0603	0.0184	0.0622	0.0189	0.0994	0.0303
0.0563	0.0171	0.0579	0.0176	0.0927	0.0282
0.0528	0.0161	0.0544	0.0166	0.0868	0.0265
0.0470	0.0143	0.0481	0.0147	0.0770	0.0235
0.0423	0.0129	0.0434	0.0132	0.0695	0.0212
0.0338	0.0103	0.0347	0.0106	0.0554	0.0169
0.02814	0.00858	0.02814	0.00883	0.0464	0.0141
0.02410	0.00735	0.02410	0.00756	0.0397	0.0121
0.02109	0.00643	0.02109	0.00662	0.0348	0.0106

3. Conductors with compact and compressed stranding have about 9 percent and 3 percent, respectively, smaller bare conductor diameters than those shown. See Table 5A for actual compact cable dimensions.
4. The IACS conductivities used: bare copper = 100%, aluminum = 61%.
5. Class B stranding is listed as well as solid for some sizes. Its overall diameter and area is that of its circumscribing circle.

Source: NEC® Handbook, NFPA, Quincy, MA, 2008, Table 8

Sizing Motor Equipment Branch-Circuit Conductors

To size a copper THWN branch-circuit conductor for a 15-hp, 208-volt, 3-phase motor:

1. Refer to *NEC* Table 430.250, which lists 3-phase motor types by horsepower, operating voltages, and full-load amperages:

   The full-load amperage for this motor is 46.2 amperes.

2. Multiply the full-load amperage value by 1.25 (125 percent) [430.22(A)]:

$$46.2 \text{ amperes} \times 1.25 = 57.75 \text{ A}$$

   The conductors for this motor must be based on an amperage value of 57.75 A.

3. Refer to *NEC* Table 310.16, which lists the allowable ampacities for conductors by material type, American Wire Gauge (AWG), circular mills, and conductor insulation type.

Answer: The minimum size Type THWN copper wire rated for 57.75 A is size 6 AWG.

Motor Branch-Circuit Short-Circuit and Ground-Fault Protection Devices

All conductors, including motor circuit conductors, must be protected from **short circuits** and **ground faults**. Short-circuit and ground-fault protective devices (e.g., fuses and circuit breakers) are placed to protect the motor and, more important, protect the conductors supplying the equipment. Motor types and the maximum size ratings allowed for the motor short-circuit and ground-fault protective devices are listed in **TABLE 3-3**, which is *NEC* Table 430.52. This table lists four types of protective devices (see **FIGURE 3-5**):

1. Non–time delay fuse (single-element, fast-acting protective device)
2. Dual-element fuse (time delay protective device)
3. Instantaneous trip breaker (fast-acting protective device)
4. Inverse time breaker (time delay protective device)

For each type of protective device, the maximum allowable size for the overcurrent device is based on a percentage of the motor's full-load amperage. Also note that when determining the size of the protective device based on calculations, the calculated values most likely do not align with standard manufactured settings for the devices. As per the *NEC*, when calculated values do not correspond to standard manufactured settings, the next higher standard can be used [Table 430.52 Exception #1 and Article 240.6].

For most general motor applications that use fuse-based protection, a dual-element time delay fuse is used, and the ratings for this fuse are 175 percent of the motor's full-load current. When circuit breakers are used to provide short-circuit and ground-fault protection (such as with panelboards), the allowable rating can be up to 250 percent of the motor's full-load current. Circuit breakers are allowed to be set at a higher value than fuses are because they are more subject to tripping during motor starts. Some manufacturers recommend that for both fuse and circuit breaker protection, the maximum size of overcurrent protection should be set at 175 percent of the motor's full-load current.

For motors that have starting currents greater than their full-load current, protective devices are allowed to be sized at values greater than the motor's actual full-load current. This sizing allows the motor to pass through the starting process and eliminates the possibility of nuisance tripping. Non–time delay fuses may be rated at up to 300 percent of the motor's full-load current.

Motor Branch-Circuit Overload Devices

Motor overload devices installed in the motor controller provide protection from overload (excessive motor running current) (see **FIGURE 3-6**). Motor overload devices operate on the principle of heat caused by the current in the circuit path and are not rated based on current alone as a method to open a circuit. Therefore, motor overload devices can be sized at a value much closer to the motor's actual full-load current, which provides for greater protection of the motor. For most applications, the

Adjusting the Size of Branch-Circuit Conductors for Voltage Drop

To adjust the size of branch-circuit conductors to account for voltage drop for the same 15-hp, 3-phase, 208-volt motor used in the previous example, which is located 275 ft from the source:

1. Determine the required circular mil area of the conductor using the following information:
   - The constant K value of the material type for the conductor: copper, 12.9
   - The length of the conductor from the source to the load: 275 ft
   - The motor's full-load amperage value (found in *NEC* Table 430.250): 46.2 A
   - The allowable voltage drop for the branch circuit (calculated by multiplying the source voltage by the 3 percent allowable voltage drop): 208 volts × 0.03 = 6.24 volts

2. Perform the following calculation:

$$Cm = \frac{K \times D \times I \times 1.73}{Vd}$$

$$Cm = \frac{12.9 \times 275 \times 46.2 \text{ A} \times 1.73}{6.24 \text{ V}}$$

Answer: 45,439 circular mil

3. Reference *NEC* Table 8, Conductor Properties, to determine the minimum size of conductor that is required.

Answer: Because the wire size must meet or exceed the calculated circular mil area to satisfy the voltage drop recommendations of 3 percent of the conductor size, the conductor size must be increased from its original size of size 6 AWG to size 3 AWG because of the distance of 275 ft.

TABLE 3-3 *NEC* Table 430.52 Maximum Rating or Setting of Motor Branch-Circuit Short-Circuit and Ground-Fault Protective Devices

Type of Motor	Percentage of Full-Load Current			
	Nontime Delay Fuse[1]	Dual Element (Time-Delay) Fuse[1]	Instantaneous Trip Breaker	Inverse Time Breaker[2]
Single-phase motors	300	175	800	250
AC polyphase motors other than wound-rotor	300	175	800	250
Squirrel cage—other than Design B energy-efficient	300	175	800	250
Design B energy-efficient	300	175	1100	250
Synchronous[3]	300	175	800	250
Wound-rotor	150	150	800	150
Direct current (constant voltage)	150	150	250	150

Note: For certain exceptions to the values specified, see 430.54.
[1]The values in the Nontime Delay Fuse column apply to Time-Delay Class CC fuses.
[2]The values given in the last column also cover the ratings of nonadjustable inverse time types of circuit breakers that may be modified as in 430.52(C)(1), Exception No. 1 and No. 2.
[3]Synchronous motors of the low-torque, low-speed type (usually 450 rpm or lower), such as are used to drive reciprocating compressors, pumps, and so forth, that start unloaded, do not require a fuse rating or circuit-breaker setting in excess of 200 percent of full-load current.

Source: *NEC® Handbook*, NFPA, Quincy, MA, 2008, Table 430.52

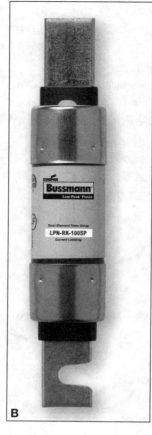

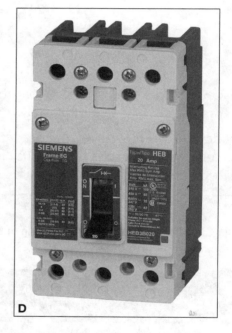

FIGURE 3-5 Protective Devices **A.** Non–time delay fuse. **B.** Dual-element fuse. **C.** Instantaneous trip breaker. **D.** Inverse time breaker.

NEC requires that motor overloads be sized at 125 percent of the motor's full-load current [430.32].

Grounding of Motors and Equipment

Typical motor applications are installed using many different types of raceway systems that are approved as an equipment grounding conductor and all must be grounded based [250]. For example, when electrical metallic tubing (Type EMT) is used, no additional equipment grounding conductor is required because Type EMT is approved as a grounding conductor.

However, with motor circuits, almost all applications have maintenance and servicing needs that require that the final connection to the motor be a more flexible type of raceway. With a flexible conduit, motor vibration or servicing of the equipment can loosen the connection at the motor termination housing, possibly eliminating any equipment grounding accomplished solely by the

Sizing Short-Circuit and Ground-Fault Protection Devices

To size short-circuit and ground-fault protection devices for a 15-hp, 3-phase, 208-volt motor:

1. Determine the full-load current of the motor by referencing *NEC* Table 430.250:

   46.2 A

2. Determine whether the device uses fuses or circuit breakers:

   A. If the motor is protected with a fuse-based device, multiply the full-load current by 1.75 (to determine 175 percent):

   $46.2 \text{ A} \times 1.75 = 80.85 \text{ A}$

Answer: For this application, *NEC* 240.6(A) *Standard Ampere Ratings* lists the closest standard sizes as 80 A and 90 A. After applying Exception #1 from *NEC* 430.52, the allowable size for a time delay fuse is 90 A.

   B. If the motor is protected with a circuit breaker–based device, multiply the full-load current by 2.5 (to determine 250 percent):

   $46.2 \text{ A} \times 2.5 = 115.5 \text{ A}$

Answer: For this application, *NEC* 240.6(A) lists the closest standard sizes as 110 A and 125 A. After applying Exception #1 from *NEC* 430.52, the allowable size for an inverse time circuit breaker is 125 A.

---

raceway system. For this reason, most motor circuit applications with flexible connections require an additional equipment grounding conductor be installed. Some electrical designers install additional equipment grounding conductors the entire length of the raceway regardless of the type of raceway system used. Although not required, this design, when installed properly, provides a more reliable grounding system.

To size equipment grounding conductors, designers must determine the value of the short-circuit and ground-fault protection devices (outlined in the previous section), and then reference **TABLE 3-4**, which is *NEC* Table 250.122, to find the minimum size for equipment grounding conductors.

When it is determined that branch-circuit conductors must be increased in size because of voltage drop concerns, the size of equipment grounding conductors must be increased in size as well. According to the *NEC*, the circular mil area of the equipment grounding conductors must be increased proportionally to the increase in the circular mil area of the ungrounded circuit conductors [250.122(B)]. An increase in the circular mil area of the equipment grounding conductors decreases resistance, thus allowing any short or fault currents to dissipate quickly from the system, avoiding damage to equipment and personnel (see "Adjusting the Size of Equipment Grounding Conductors for Voltage Drop" on page 39).

The added safety of an equipment grounding conductor along the entire length of the raceway often outweighs the additional material costs.

In any application where branch-circuit conductors are increased in size, the size of equipment grounding conductors must be increased proportionally.

FIGURE 3-6 Motor overload device in motor controller.

TABLE 3-4 *NEC* Table 250.122 Minimum Size Equipment Grounding Conductors for Grounding Raceway and Equipment

Rating or Setting of Automatic Overcurrent Device in Circuit Ahead of Equipment, Conduit, etc., Not Exceeding (Amperes)	Size (AWG or kcmil) Copper	Aluminum or Copper-Clad Aluminum*
15	14	12
20	12	10
30	10	8
40	10	8
60	10	8
100	8	6
200	6	4
300	4	2
400	3	1
500	2	1/0
600	1	2/0
800	1/0	3/0
1000	2/0	4/0
1200	3/0	250
1600	4/0	350
2000	250	400
2500	350	600
3000	400	600
4000	500	800
5000	700	1200
6000	800	1200

Note: Where necessary to comply with 250.4(A)(5) or (B)(4), the equipment grounding conductor shall be sized larger than given in this table. *See installation restrictions in 250.120.

Source: *NEC® Handbook,* NFPA, Quincy, MA, 2008, Table 250.122

■ Raceways for Branch-Circuit Distribution

Once all the branch circuit and equipment grounding conductor sizes have been determined, the next step is to determine the raceway type and sizes that will be used.

The *NEC* provides many approved raceway methods for branch circuits to supply loads and equipment. When the electrical wiring to be installed is in an existing facility, the conduit and raceway system most likely will be an EMT raceway. EMT systems are easily installed and durable but may be more costly than other methods. In new buildings and where possible in other projects, installing conduits underground using rigid nonmetallic conduit (PVC) can help reduce costs.

The following *NEC* tables are valuable references for designing raceways for branch-circuit distribution:

Annex C Conduit and Tubing Fill Tables for Conductors and Fixture Wires of the Same Size

Table 4, Chapter 9 Dimensions and Percent Area of Conduit and Tubing

Table 5, Chapter 9 Dimensions of Insulated Conductors and Fixture Wires

Adjusting the Size of Equipment Grounding Conductors for Voltage Drop

To adjust the size of equipment grounding conductors to account for voltage drop for a 15-hp, 3-phase, 208-volt motor located at a distance of 275 ft from the source:

1. Determine the size of the branch circuit conductors for a normal circuit:

    26,240 cm (size 6 AWG)

    (See "Sizing Motor Equipment Branch-Circuit Conductors" earlier in this chapter.)
2. Determine the size of the branch-circuit conductors after adjusting for voltage drop:

    45,439 cm (size 3 AWG)

    (See "Adjusting the Size of Branch-Circuit Conductors for Voltage Drop" earlier in this chapter.)
3. Calculate the percentage of increase by dividing the adjusted size by the normal size:

    $$\frac{45,439 \text{ cm}}{26,240 \text{ cm}} = 1.73 \times 100 = 173 \text{ percent}$$

4. Determine the size of the equipment grounding conductor by determining the value of the short-circuit and ground-fault devices and referencing *NEC* Table 250.122 (before making any adjustments for voltage drop):

    size 8 AWG = 16,510 cm

    (See "Sizing Short-Circuit and Ground-Fault Protection Devices" earlier in this chapter.)
5. Calculate the required increase of the equipment grounding conductors' size by multiplying the size by the same percentage:

    16,510 cm × 1.73 = 28,562 cm

6. Reference *NEC* Table 8 to determine the adjusted size for equipment grounding conductors: 28,562 cm = size 4 AWG

Answer: This equipment grounding conductor should be resized as a size 4 AWG conductor.

In raceway systems, individual conduits can be installed from the distribution source to an individual piece of equipment, but often a single conduit of a larger size that contains multiple branch circuits for more than one piece of equipment is used. With this method, single, larger conduits are installed in close proximity to the equipment and terminated in properly sized junction or pull boxes. From these junction or pull boxes, smaller individual conduits are run to each individual piece of equipment (see **FIGURE 3-7**).

Regardless of the type of raceway system, the conduits must be properly sized and designed for the amount of conductors. To calculate the proper size of the raceway, the designer must determine whether the conductors to be installed in a raceway are all of the same circular mil or AWG size and insulation type. If they are the same size and type, the designer can choose the raceway size using the tables in *NEC* Annex C, which lists raceway types by trade size in both metric and standard fractional sizes and the allowable number of conductors permitted. If the conductors are not of the same size and insulation type, the designer can determine the total area of the conductors and reference *NEC* Chapter 9 Table 4 and Table 5.

For conductors of different AWG/kcmil sizes and insulation types, *NEC* Chapter 9 Table 4 provides a great deal of information about each raceway type in both metric and standard measurements such as size, internal diameters, total area for its given size, and the allowable areas that may be used when multiple types of conductors are installed. The information in Table 4 is based on calculations of the area

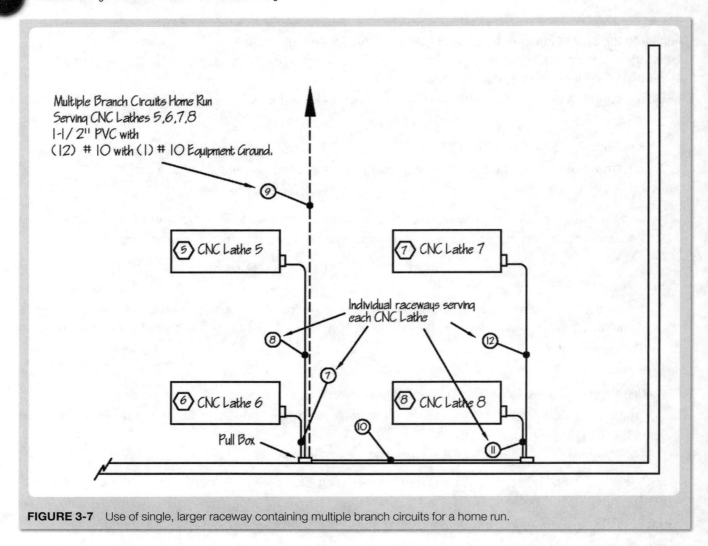

FIGURE 3-7 Use of single, larger raceway containing multiple branch circuits for a home run.

each conductor will fill in the raceway; therefore, all conductors, including any required equipment grounding conductors, must be counted as part of the total area of the conductors. Because conductors are available with many different insulation types of varying thicknesses, conductors have variable diameters and total square inch area. *NEC* Chapter 9 Table 5 provides information about conductor types, including insulation type, size in AWG or kcmil, approximate diameter, and approximate area. This table should be used to calculate the total area of conductors to be installed in any raceway type.

When determining the maximum conduit fill for an installation, it is permissible to increase a conduit size. Using a raceway larger than the minimum required size may lead to additional cost, but if the conduit run is long or contains several bends, larger conduit eases pulling of the conductors, thus saving labor costs and possibly preventing damage to the conductors' insulation.

> **TIP**
> The appropriate conduit size in square inches must be more than the total calculated area for variable-sized conductors.

■ Determining the Number of Panelboards

All projects require a minimum number of panelboards based on the minimum number of required

Determining Conduit Fill for Conductors of the Same Size

To determine conduit fill for six size 4 AWG THWN conductors to be installed in an electrical metallic tubing (Type EMT) conduit:

1. Find the appropriate table in *NEC* Annex C based on conduit type:

    EMT is listed in Table C1.

2. Locate the appropriate section for Type THWN conductors:

    The seventh section down is appropriate for thermoplastic high heat-resistant nylon-coated (Type THHN), Type THWN, and Type THWN-2 conductors.

3. Locate the appropriate row for size 4 AWG conductors:

    The sixth row down is appropriate for size 4 AWG conductors.

4. Look across this row until a number higher than the total number of conductors (in this case, six) is found:

    The fourth column in this row is appropriate for up to seven conductors.

5. Look at the top of this column to find the minimum required conduit size for this type of conduit:

    The minimum size for this conduit is 1¼ in.

Partial Values Table C1 National Electrical Code

TYPE	CONDUCTOR SIZE (AWG/kcmil)	CONDUCTORS TRADE SIZE				
	(Metric)	16	21	27	35	41
		1/2"	3/4"	1"	1-1/4"	1-1/2"
THHN THWN THWN-2	14	12	22	35	61	84
	12	9	16	26	45	61
	10	5	10	16	28	38
	8	3	6	9	16	22
	6	2	4	7	12	16
	4	1	2	4	7	10
	3	1	1	3	6	8
	2	1	1	3	5	7
	1	1	1	1	4	5

Excerpt from *NEC* Table C.1 Maximum Number of Conductors or Fixture Wires in Electrical Metallic Tubing (EMT)
Source: *NEC® Handbook*, NFPA, Quincy, MA, 2008, Table C1

branch circuits and the operating voltages of the loads to be served. Panelboards typically have a maximum number of 42 branch-circuit spaces available, but as per changes in the 2008 version of the *NEC*, the number of allowed branch circuits in any one panelboard is no longer limited to 42. Panelboards can contain as many branch circuits as listed for specific panelboards. According to the *NEC*:

A panelboard shall be provided with physical means to prevent the installation of more overcurrent devices than that number for which the panelboard was designed, rated, and listed. For the purposes of this section, a 2-pole circuit breaker or fusible switch shall be considered two overcurrent devices; a 3-pole circuit breaker or fusible switch shall be considered three overcurrent devices [408.54].

With large quantities of single-phase loads requiring one or two poles and 3-phase loads requiring three poles, the 42 individual branch-circuit spaces in a single panelboard can quickly be used. If the required number of branch circuits for the equipment at their serving voltages exceeds 42, then an additional panelboard may be necessary.

Determining Conduit Fill for Conductors of Variable Size

To determine conduit fill for a Type EMT conduit with the following conductors:
- A. Four size 10 Type THWN conductors
- B. Three size 6 Type THWN conductors
- C. Three size 4 Type THWN conductors

1. Determine the total area for each of the individual conductors.
2. Multiply each of the conductor's individual area by the quantity to be installed (see *NEC* Table 5 for conductor area):
   - A. 4×0.0211 in.^{2} = 0.0844 in.^{2}
   - B. 3×0.0507 in.^{2} = 0.1521 in.^{2}
   - C. 3×0.0824 in.^{2} = 0.2472 in.^{2}
3. Add these totals to find the total area for all conductors:

   0.0844 in.^{2}
   0.1521 in.^{2}
   $+ 0.2472$ in.^{2}
   _____
   0.4837 in.^{2}

4. Refer to *NEC* Table 4, Article 358, to find the appropriate conduit type based on the total area:

   In the Over Two Wires 40% column, see that a 1-in. Type EMT conduit has a maximum fill of 0.346 in.^{2} and a 1¼-in. Type EMT conduit has a maximum fill of 0.598 in.^{2}. Because the calculation total of 0.4837 in.^{2} in step 3 is between these two allowed areas, the larger of the two values must be used.

Answer: A 1¼-in. conduit should be used for these conductors.

To determine the minimum number of panelboards, designers consider the following factors:
- The operating voltage of the loads to be served
- Number of branch-circuit requirements
- Size of the facility
- Location of the equipment in relationship to the panelboards
- Horsepower rating of equipment to be served
- Labor and material costs associated with each installed panelboard

A good design incorporates all the minimum requirements and provides an efficient system to serve the facility safely with an ability to expand for future needs.

Larger Motors

Motors 25 hp and greater require larger amounts of amperage than do motors of lower horsepower. If these loads are designed to be served from a branch circuit panelboard, the overall amperage capacity of the panelboard must be increased. To do so, designers can increase the amperage size of the panelboard, the feeder conductors, and conduit systems that serve these panelboards to accommodate the additional current and electrical stress. Labor and material costs also increase with the increase in amperage capacity of the panelboard. In such cases, it may be advisable to consider an alternative design method serving the loads directly from the main switchboard and eliminating a larger ampacity panelboard and associated costs.

Equipment Branch Circuits and Panel Schedules

Once a determination has been made as to which equipment branch circuits will be served from a particular panelboard, the designer should arrange the branch circuits within the panelboard. Just as with general purpose branch circuits (see Chapter 2), the designer enters the specialty branch circuits into a panel schedule. The process is very similar to the one used for general purpose equipment except that specialized equipment may contain both single- and 3-phase motors.

(See "You Are the Designer" at the end of this chapter for an example of how to design a panelboard in this situation.)

For each specialty branch circuit, the designer must calculate a total volt-ampere rating using Ohm's law. This calculation is based on the equipment's voltage and amperage and determines the volt-ampere values to be entered into the panel schedule. The total volt-amperes of the equipment is divided by the appropriate number of ungrounded conductors required for the load.

■ Raceway Legends

<u>Raceway legends</u> in the design plan provide a great deal of information about individual raceways and conductors for special equipment loads (see **FIGURE 3-8**). Although no official industry-standard format exists, a raceway legend generally lists the following information:

- Raceway I.D.
- Quantity and size of conductors
- Equipment grounding conductor size
- Raceway size
- Raceway type
- Panelboard serving load
- General remarks

A raceway legend provides clear and concise information in a simple format and helps to avoid excess information on the electrical plan that might otherwise cause confusion, errors, or omissions. Although it can take time to develop a raceway legend and this legend is not a requirement, including a raceway legend provides a higher quality design and demonstrates a greater degree of professionalism.

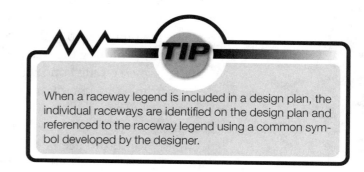

TIP: When a raceway legend is included in a design plan, the individual raceways are identified on the design plan and referenced to the raceway legend using a common symbol developed by the designer.

I.D.	Qty. / Size of Conductors	Equip. Gr. Conductor	Raceway Size	Raceway Type	Panelboard	Remarks
1	(3) #10	(1) #10	1"	EMT	P2'	CNC Lathe #1
2	(3) #10	(1) #10	1"	EMT	P2'	CNC Lathe #2
3	(6) #10	(1) #10	1-1/4"	EMT	P2'	CNC Lathe #1 & 2
4	(3) #10	(1) #10	1"	EMT	P2'	CNC Lathe #4
5	(3) #10	(1) #10	1"	EMT	P2'	CNC Lathe #3
6	(12) #10	(1) #10	1-1/2"	PVC	P2'	CNC Lathe #1,2,3,4, Home Run
7	(3) #10	(1) #10	1"	EMT	P1'	CNC Lathe #6
8	(3) #10	(1) #10	1"	EMT	P1'	CNC Lathe #5
9	(12) #10	(1) #10	1-1/2"	PVC	P1'	CNC Lathe #5,6,7,8 Home Run
10	(6) #10	(1) #10	1-1/4"	EMT	P1'	CNC lathe #7,8
11	(3) #10	(1) #10	1"	EMT	P1'	CNC Lathe #8
12	(3) #10	(1) #10	1"	EMT	P1'	CNC lathe #7

FIGURE 3-8 The raceway legend provides information about the raceways and conductors that serve specialized equipment loads.

Wrap Up

■ Master Concepts

- Equipment lists outlining all the equipment in a facility are necessary to determine the number of branch circuits needed and the quantity and operating voltages of panelboards.
- Because of their amperage requirements and manufacturer specifications, specialized equipment and motorized equipment must be served by their own individual branch circuits.
- When loads are located farther than 100 ft from their source, because of the resistance of the conductors serving specialized and motorized equipment loads, the circular mil area of branch circuit conductors must be increased to provide for a maximum voltage drop of 3 percent of the source voltage.
- Properly determining the size of the overcurrent protection for motors, equipment, and the branch circuit conductors that serve them is very important to protect these devices from short-circuit or ground-fault conditions.
- When designing panelboards, designers should aim to meet all the minimum requirements and provide an efficient system to serve the facility safely with an ability to expand for future needs.
- A raceway legend, though not a requirement, is an extremely useful tool to identify the raceways and the conductors contained within them for identification on a design plan.

■ Charged Terms

277/480-volt, 3-phase, 4-wire, wye system A distribution system generated with three individual sine waves separated by 120 electrical degrees that are identified as phases A, B, and C. One leg of each of the three phase coils is electrically connected to the others at a common point, forming a wye, which when grounded becomes the fourth wire (or neutral) in the system. This allows for each of the three individual phase voltages to supply 277 volts to the grounded point, while the line voltage across each of the phases produces 480 volts. The line-to-line voltages can supply both 480-volt 3-phase and 480-volt single-phase. This system is commonly used in commercial applications where 480 volts is required for machinery loads and in applications to serve 277-volt lighting loads.

equipment grounding conductor The conductive path installed to connect normally non-current-carrying metal parts of equipment together and to the system grounded conductor, to the grounding electrode conductor, or to both [100].

equipment list A developed table that lists details about the specialized equipment that is to be incorporated into a design plan.

ground fault A condition in which high levels of current could flow when an ungrounded conductor accidentally comes in contact with a grounded reference.

inrush current A momentary high level of amperage flowing in a circuit such as those associated with motorized equipment loads.

isolated ground (IG) An additional equipment grounding conductor that, when installed, provides for the grounding of equipment separate from a grounding method that uses an approved raceway method; typically used for electrically sensitive equipment in computer applications and medical facilities. When isolated grounding is provided through the use of receptacles, the receptacle must be identified on the design plan as "IG."

manufacturer electrical specification sheet Information provided by a manufacturer that lists specific details about the product; these specification sheets are often used to obtain information about motorized equipment and lighting fixtures.

raceway legend A table developed by an electrical designer that illustrates information about raceways installed for a project.

short circuit A dangerous condition in which circuit conductors contact each other and reduce the intended ohmic resistance of the circuit; often referred to as a line-to-line or line-to-neutral short. (*See also* ground fault.)

voltage drop A loss of voltage on a conductor resulting from the length of the conductor, its resistance, and the amperage imposed on the conductor.

■ Check Your Knowledge

1. The purpose of an equipment list is to:
    A. list the electrical requirements for any equipment to be installed on a project.
    B. provide manufacturer specification sheets to the owner.
    C. assign each equipment piece an identification number.
    D. assign each equipment piece a branch circuit number.
2. A raceway legend provides:
    A. the types and quantities of raceway material to be used on a project.
    B. a detailed layout for the routing of all the raceways for a project.
    C. a list of raceways by type, size, and the conductors to serve equipment loads.
    D. a listing of all the raceway types for a project.
3. A 10-hp, 208-volt, 3-phase motor located 50 ft from the source requires a size _____ AWG copper Type THWN conductor.
    A. 12
    B. 10
    C. 8
    D. 6

4. A 10-hp, 208-volt, 3-phase motor located 350 ft from the source requires a size _____ AWG copper Type THWN conductor.

   A. 20
   B. 8
   C. 6
   D. 4

5. What is the maximum allowable size for a time delay fuse that provides protection for a 10-hp, 208-volt, 3-phase motor?

   A. 30 A
   B. 40 A
   C. 50 A
   D. 60 A

6. What is the maximum allowable size for an inverse time circuit breaker that provides protection for a 10-hp, 208-volt, 3-phase motor?

   A. 60 A
   B. 70 A
   C. 80 A
   D. 90 A

7. A 10-hp, 208-volt, 3-phase motor located 75 ft from the source and protected by a time delay fuse requires a size _____ copper Type THWN equipment grounding conductor.

   A. 12 AWG
   B. 10 AWG
   C. 8 AWG
   D. 6 AWG

8. A 10-hp, 208-volt, 3-phase motor protected by a time delay fuse and located 350 ft from the source requires a size _____ copper Type THWN equipment grounding conductor.

   A. 12 AWG
   B. 10 AWG
   C. 8 AWG
   D. 6 AWG

9. What is the minimum raceway size for a schedule 40 PVC conduit serving a 10-hp, 208-volt, 3-phase motor located within 100 ft of the source?

   A. ½ in.
   B. ¾ in.
   C. 1 in.
   D. 1¼ in.

10. For the motor described in question 9, what is the total volt-amperage per phase?
    A. 6,406 VA
    B. 11,083 VA
    C. 14,784 VA
    D. 25,576 VA

11. For the motor described in question 9, what is the individual volt-amperage per phase?
    A. 2,135 VA
    B. 3,694 VA
    C. 4,928 VA
    D. 8,525 VA

You Are the Designer

Apply the knowledge you have gained from this previous chapter to your own electrical design. In this section you will:
- Develop an equipment list for specialized equipment in both the general office and manufacturing areas
- Design the required circuitry and branch circuits for the equipment
- Develop panel schedules for the office specialty and motorized equipment loads
- Determine and design the raceway systems that will serve all the equipment branch circuits
- Develop a raceway legend for the raceway design that illustrates all the requirements (e.g., equipment identification, raceway type, quantity and size of the conductors, equipment grounding conductors)
- If necessary, calculate adjustments for voltage drop if any branch circuit conductors serving the equipment loads need to be increased
- Check the plan for completeness and accuracy

■ About Your Project

To complete your task, you must know the following details about your project:
- Specialized equipment that is to be served in the general office areas
- Information about motorized equipment loads that are to be served in the manufacturing area, including operating voltages, horsepower, and phases

■ Resources

To develop this part of your design, you need the following resources:
- The following documents from the *Student Resource CD-ROM*:
  - Plan sheets E_1 and E_3
  - Equipment list
  - Panel schedule for a 120/208-volt, 3-phase, 4-wire system
  - Panel schedule for a 277/480-volt, 3-phase, 4-wire system
  - Raceway legend
  - Electrical symbol list
- *NEC* Tables 430.248 and 430.250
- The panel schedule you developed in Chapter 2

■ Get to Work

Office Area Specialized Equipment Loads

In this part of the design process, you develop a list of specialized equipment that is to be served in the office areas. You may wish to select specialized equipment other than those listed in the text to enhance your understanding of the design process. If you choose to work with the values used in the completed project, review the equipment list on completed plan E_1, which is based on the following values:

Vending machine	1800 VA
Microwave	1200 VA
Copier	1440 VA
Computer network server	1200 VA

Most office specialty loads operate on 120-volt sources with maximum volt-ampere ratings of 1800 VA. To design for and create panel schedules for any load for which volt-ampere values are not provided but the equipment is rated in amperage, you must perform calculations to determine the equipment volt-ampere ratings.

Determining Volt-Ampere Values for 120-Volt Equipment

To determine the volt-ampere values for 120-volt loads when amperage values are given:

1. Obtain the amperage rating for the equipment either from the manufacturer or from *NEC* Table 430.248.
2. Determine the volt-ampere rating of the equipment using Ohm's law:
   Volt-amperes = Amperes (I) × Voltage (E)

Example: For a specialty piece of 120-volt equipment with an amperage value of 10 amperes:

$$10 \text{ A} \times 120 \text{ V} = 1200 \text{ VA}$$

Next, you need to enter these values into the panel schedule you created in the last chapter. As you enter the loads, assign them to circuit numbers so that they are balanced equally across all three of the phases as best as possible (for an example, see completed panel schedule P_3). If the loads are unbalanced, move the assigned branch circuits to alternate phases to achieve better balancing.

Manufacturing Area Equipment Loads

Specialized loads often have different voltages and numbers of phases (such as 208 volts and 480 volts). You must create a panel schedule for each of the serving voltages. Once you have selected all the motors, enter all the required information into your equipment list including the horsepower, voltage, phase, and amperage ratings, which you can obtain from manufacturer specification sheets. Then, perform any necessary calculations to obtain volt-ampere ratings for the panel schedule.

Determining the Volt-Ampere Ratings for Three-Phase Loads Operating on 208 or 480 Volts

To determine the volt-ampere values for 208- or volt, 3-phase loads when amperage values are given:

1. Obtain the operating voltage.
2. Obtain the load in amperage from the manufacturer specification sheet or *NEC* Table 430.250.

3. Use Ohm's law for 3-phase applications:
   Ohm's law:
   $$\text{Volt-amperes} = \text{Amperes}(I) \times \text{Voltage}(E) \times 1.73$$
   For example, for a 15-hp, 480-volt, 3-phase motor with an amperage of 21.0 full-load amperes:
   $$21.0 \text{ A} \times 480 \text{ V} \times 1.73 = 17{,}438 \text{ VA}$$

Note: For a 208-volt application, perform the calculation by substituting 208 volts for the 480-volt value.

Because of differing service voltages and phases, you need to create panel schedules for the appropriate voltage. First, determine the volt-ampere values for each phase (divide the total calculated volt-ampere value by the number of phases serving the load). For example, for a 3-phase motor with a volt-ampere value of 17,438 VA, divide the total by 3 to determine the volt-ampere load for each phase (17,438 / 3 = 5813 VA per phase). If the load to be served is a single-phase, 208- or 480-volt load requiring two ungrounded conductors, divide the total calculated volt-ampere rating by 2 to determine the total volt-ampere load for each of the two ungrounded conductors.

TIP

When designing panelboards, 208-V, single- and 3-phase loads and 120-V single-phase loads may be served from the same panelboard. Any 480-V single-phase and 3-phase loads must be served by an additional panelboard.

Determining the Individual Phase Volt-Ampere Values for Single and Three-Phase Equipment

To determine the individual phase volt-ampere values for single and 3-phase equipment:

1. Obtain the voltage of the equipment to be served.
2. Determine the number of ungrounded conductors necessary to serve the load (single-phase 208-volt and 480-volt loads are served with two ungrounded conductors; 3-phase loads are served with three ungrounded conductors).
3. Divide the total volt-ampere value by the number of ungrounded conductors necessary to serve the load.

   Example 1: A 480-volt, single-phase welder with a total volt-ampere rating of 50,000 VA (single-phase, two ungrounded conductors required):
   50,000 / 2 = 25,000 VA for each ungrounded conductor

   Example 2: A 480-volt, 3-phase welder with a total volt-ampere rating of 50,000 VA (3-phase load, three ungrounded conductors required):
   50,000 / 3 = 16,667 VA for each ungrounded conductor.

Once you have calculated these values, complete the required number of panel schedules for your design. If there are more than 42 branch circuits, create additional panel schedules to accommodate the additional panelboards that you will incorporate into your final design. Remember, to reduce costs, equipment of larger horsepower ratings can be served directly from the main switchboard, eliminating the need for the equipment to be served through individual panelboards (which results in larger panelboards and feeder sizes).

At this point in your panel schedule design, enter into your panel schedules the title of the loads to be served by their appropriate branch circuits (e.g., CNC #1). For each load you enter into a panel schedule, enter the appropriate volt-ampere rating for each phase that is to serve the equipment.

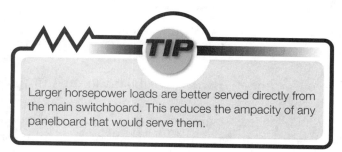

TIP

Larger horsepower loads are better served directly from the main switchboard. This reduces the ampacity of any panelboard that would serve them.

Raceway Design

The next step is to determine the raceway methods that will be used to serve all the branch circuitry for the equipment loads in the manufacturing area. Although an individual raceway may be used for each piece of equipment, it can be more cost-efficient to provide one larger raceway from the serving panelboard to a location in close proximity to groups of equipment located in a facility. This larger raceway can then terminate in a junction or pull box, from which the individual branch circuits are routed to each piece of equipment in an individual raceway.

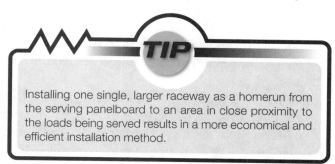

TIP

Installing one single, larger raceway as a homerun from the serving panelboard to an area in close proximity to the loads being served results in a more economical and efficient installation method.

This method is illustrated in the completed power plan on sheet E_1. Notice how the branch circuits for Grinders 9, 10, and 11 are initially installed in a 1¼-in. conduit terminated in a junction box (see raceway legend #19) and how individual conduits were then run to each individual grinder from the junction box termination point of the 1¼-in. conduit. This eliminates the need to install three individual raceways from panelboard P_2 to each grinder and achieves more efficient use of materials and labor.

Because this facility is new construction, the designer is allowed the option of using rigid nonmetallic type (PVC) conduit installed underground (before concrete floors are installed) for the raceways for the branch circuit runs to the junction boxes. Rigid nonmetallic raceways are far less costly than raceways installed aboveground. Aboveground methods require the use of metallic type raceways, such as Type EMT, and metallic raceway support methods, the material and labor costs of which far exceed those of PVC type raceway methods. If a project is an alteration (remodel) and there is no access under an existing concrete floor, underground PVC type raceway methods may not be an option.

The use of an underground non-metallic raceway method, such as PVC, is a more economical method of branch circuit distribution.

The completed design includes three panelboards:
- 277/480-volt, 3-phase, 4-wire panelboard located in the manufacturing area (P_1)
- 120/208-volt, 3-phase, 4-wire panelboard located in the manufacturing area (P_2)
- 120/208-volt, 3-phase, 4-wire panelboard located in the office area (P_3)

For your project, you can design your own panelboard distribution, adding or deleting panelboards as necessary. But remember, any panelboard installed in the general office area should serve electrical loads only for that area and should not include any equipment loads located in the manufacturing area.

To design the raceways for your project, you must determine which pieces of equipment are to be served from panelboards and which, if any, will be served directly from the main switchboard. For this part of the design, the sizing of the raceways is not required. This step illustrates the raceway design and, more important, enables you to develop a measured length of the raceways so that you can determine the lengths and proper sizes of the branch circuit conductors.

Start by determining which type of approved raceway method(s) you will use and how the raceways will be installed. Illustrate all the raceways on your plan to the appropriate distribution point and use the appropriate symbols for each. For example, if you choose to design a raceway system that will be installed underground, illustrate it by using a dashed line.

Once you complete the raceway design, you must calculate the required AWG/kcmil sizes for each piece of equipment based on the voltage and horsepower values in the equipment list. Use the methods discussed earlier in this chapter to size conductors correctly. Remember that for any branch circuit that is installed to serve equipment, you need to perform voltage drop calculations if the distance from the source exceeds 100 ft. Ensure that the conductors are of adequate AWG/kcmil size, limiting the maximum voltage drop to not more than 3 percent of the source voltage.

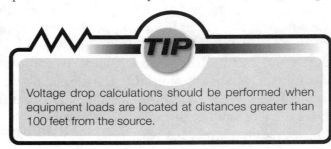

Voltage drop calculations should be performed when equipment loads are located at distances greater than 100 feet from the source.

With all the branch circuit conductor sizes determined, you can now calculate the proper size of the raceways. From your raceway design, determine how many conductors and their AWG/kcmil sizes are to be installed in each of the raceways based on the equipment you have designed for. If your design incorporates multiple branch circuits in a single raceway, the *NEC* requires only one equipment grounding conductor in the raceway [250.122(C)]. To size the equipment grounding properly you must determine the size of the largest overcurrent protective device protecting the conductors in the raceway, and then use *NEC* Table 250.122 to reference the correct AWG/kcmil size for the equipment grounding conductor.

Determining the Size of Equipment Grounding Conductors for Multiple Branch Circuits Enclosed in the Same Raceway

To size an equipment grounding conductor properly when multiple branch circuits are enclosed in the same raceway:

1. Determine the size of the largest branch circuit overcurrent device serving the conductors in the raceway (for motor and equipment circuits, this is generally the overcurrent protective device of the largest motor).

2. Use *NEC* Table 250.122 to determine the correct equipment grounding conductor size. (Remember, if the branch circuit conductors have been increased in AWG/kcmil size because of voltage drop concerns, the equipment grounding conductor size must be increased proportionally as well.)

Example: A conduit has multiple copper branch circuits in a raceway. The size of the largest overcurrent protection for the branch circuits is 60 A. Therefore, the minimum size copper equipment grounding conductor must be size 10 AWG.

Finally, complete a raceway legend for all the raceways you have designed using the raceway legend on the *Student Resource CD-ROM* and the completed raceway legend on Completed Plan Sheet E_1.

CHAPTER 4

Lighting Systems

Chapter Outline

- Introduction
- Determining Lighting Requirements
- Selecting Lighting Fixtures
- Calculating the Number of Lighting Fixtures Required
- Determining Lighting Fixture Location
- Creating the Lighting Plan
- Including Keynotes Sections
- Complying with Energy Code Requirements
- Interpreting Energy Code Design Methods

Learning Objectives

- Analyze and evaluate lighting design standards to determine the lighting requirements of a facility.
- Apply the appropriate concepts to determine the quantity of lighting fixtures necessary to meet a facility's lighting needs.
- Determine optimal quantities and locations of lighting fixtures to provide sufficient lighting.
- Develop a lighting fixture schedule.
- Recognize the various components of a lighting plan including branch circuits, control devices, and raceways.
- Synthesize knowledge of regulatory energy codes into a lighting design plan to meet energy allowance requirements using appropriate design methods.

Introduction

Lighting systems are one of the most important components of any facility. **Lighting systems** include lighting fixtures and components that operate and control them. In today's commercial facilities, controls may be simple switches on the wall or more advanced control devices that allow lighting operation to be programmed for several on and off sequences, depending on hours of operation.

The goal of lighting design is to develop a high-quality lighting system that meets a facility's lighting needs while adhering to strict mandated energy codes that limit the amount of power a facility can use for the lighting system. A lighting system that complies with these energy codes while still providing a customer with a high-quality design is often the most detailed component of a design plan. This stage of design has several parts. First, the electrical designer must determine the facility's lighting requirements, select appropriate lighting fixtures, and calculate the necessary number of lighting fixtures to meet lighting requirements. Then, the designer must determine the spacing of lighting fixtures throughout the facility and create the lighting design, noting all the branch circuits, raceways, switches, and sensors necessary to meet lighting and energy code requirements.

Determining Lighting Requirements

For a commercial lighting design to be successful, each area of the building must have a sufficient level of light to allow users to perform their tasks. Each facility has unique lighting requirements based on the tasks performed in that area (e.g., manufacturing or general office tasks). In addition to general lighting throughout the facility, the designer must consider the specific requirements of each area, such as lighting to reduce screen glare in areas with computer monitors. Commercial buildings may also have showrooms that require a specialized lighting design to feature products for sale.

In the United States, light luminance levels are typically measured in **foot candles (fc)** though occasionally the International System of Units (SI) measurement—lux—is used. The Illuminating and Engineering Society of North America (IESNA), an association of engineers and lighting design professionals who devote their time to the study and development of lighting, publishes the *IESNA Lighting Handbook,* which provides recommended **luminance levels** for all types of occupancies. For most general office areas, lighting luminance levels typically range from 30 to 50 fc (see **FIGURE 4-1**). Luminance levels in manufacturing areas can be as low as 20 fc or as high as 300 fc, based on the precision of tasks to be performed in that area (see **TABLE 4-1** and **FIGURE 4-2**).

FIGURE 4-1 Commercial areas typically require lower luminance levels.

The Illuminating and Engineering Society of North America (IESNA) provides valuable documentation for lighting design criteria and recommendations.

TABLE 4-1 Recommended Lighting Foot Candle Levels

Area	Recommended Lighting Level
Office Areas	
General Spaces	30 fc
Detailed Paperwork	50 fc
Manufacturing Areas	
Rough Work	20 fc
Detail Work	50 fc
Fine Detail Work	100 fc
Very Fine Detail Work	300 fc

FIGURE 4-2 Some manufacturing areas require high-level lighting.

Selecting Lighting Fixtures

To select appropriate lighting fixtures, the designer must consider the following factors:
- Design (aesthetics)
- Efficiency
- Lamp type
- Mounting method

Manufacturers in the lighting industry are always developing newer lighting technologies. Designers should obtain current lighting fixture catalogs and accompanying manufacturer specification sheets during the design phase to determine these criteria.

Designers must consider the lighting fixture's ratio between output (luminance) and input (power). This ratio is measured in **lumens per watt (lm/W)** (a lumen is a measure of total light output of the lamp). The less input energy that is required for light output, the higher the lumen per watt rating. Typical incandescent lamps have lumens per watt ratios of approximately 10 to 20 lm/W; newer compact fluorescent lamps have lumens per watt ratios of approximately 40 to 70 lm/W. Fluorescent lamps have a lumen output of approximately four times that of incandescent lamps for the same amount of input power.

The lumens per watt ratio affects the quality of light generated by the lamp. The quality (or color) of lighting is known as the **color rendering index (CRI)** rating. CRI measures the percentage to which the color of an object viewed by a particular light source is similar to the color of that object viewed in natural daylight. Because true color rendition is not essential in most commercial facilities, lighting fixtures with fluorescent lamps are commonly used. Fluorescent lamps have a CRI value of 84 to 90 and come in many styles and design types. They deliver a high-quality light with relatively low fixture costs and levels of energy consumption. However, in retail areas or showrooms where true color rendition of products is important, halogen or metal halide lighting fixtures are more common. These types of fixtures provide optimal lighting for viewing true colors with CRI ratings near 100, and they are available with high lumens per watt ratios of about 65 to 155 lm/W.

Calculating the Number of Lighting Fixtures Required

Once the lighting fixtures have been chosen for the application, the designer can calculate the number required to meet the facility's lighting requirements. Several variables factor into this calculation:
- Lighting requirements for the space measured in foot candles (fc)
- Total area (A) of the space to be illuminated, in square feet
- Fixture lumens (FL)—total light output, provided by the lamp manufacturer
- Light loss factor (LLF)—derating factor, of the light output that reduces the overall light output over time from dirt accumulation on the lamp and fixture lens
- Coefficient of utilization (Cu)—a measure of how efficiently light produced within a lighting fixture is distributed across the area to be illuminated

Some sample values for fixture lumens can be found in **TABLE 4-2**. Light loss factor can vary depending on maintenance cleaning cycles and how clean the operating environment of the lighting fixture is. The coefficient of utilization is determined by the manufacturer and is dependent upon how well the lumen output from the lamp is distributed from the lighting fixture. Not all light produced by the lamp within the

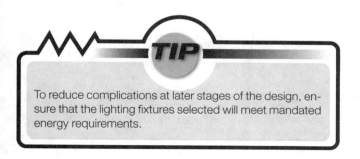

To reduce complications at later stages of the design, ensure that the lighting fixtures selected will meet mandated energy requirements.

lighting fixture actually exits the lighting fixture and reaches the surface to be illuminated.

TABLE 4-2 Lamp Lumens

Wattage	Lamp Type	Fixture Lumens per Lamp
32 watts	Fluorescent	2,850 lumens
26 watts	Compact fluorescent	1,800 lumens
32 watts	Compact fluorescent	2,400 lumens
42 watts	Compact fluorescent	3,200 lumens
400 watts	Metal halide (HID)	36,000 lumens
1000 watts	Metal halide (HID)	110,000 lumens
54 watts	Fluorescent T5 (high output)	4,400 lumens

Quality of the lighting fixture (such as design, reflectivity of the paint surfaces within the lighting fixture, etc.) determines how well the light produced within the lighting fixture actually exits it.

It can be very time-consuming to perform lighting fixture calculations manually. Special lighting design software is an extremely valuable tool for completing these calculations because it calculates the quantity of lighting fixtures required, generates drawings illustrating the luminance levels for all areas of the room based on the mounting locations for the lighting fixtures, and quickly evaluates alternative lighting fixture types.

Typically, designers calculate the number of fixtures required using lighting design software. However, designers can use the following equation to calculate basic lighting needs manually:

$$\text{Number of lighting fixtures required} = \frac{fc \times A}{FL \times LLF \times Cu}$$

Calculating the Number of Lighting Fixtures Required

To calculate the number of fixtures required for an office with the following specifications:
- Dimensions of 25.5 ft × 16.5 ft
- Desired luminance level of 30 fc
- Fluorescent lighting fixture with three fluorescent lamps, each with an output of 2850 lm

1. Calculate the total area of the space to be illuminated (A):

$$25.5 \text{ ft} \times 16.5 \text{ ft} = 420.75 \text{ ft}^2$$

2. Determine the total lumen output of the lamp(s) (FL):

$$\text{Number of lamps within the lighting fixture} \times \text{Lumen output per each lamp} = FL$$

$$3 \text{ lamps} \times 2850 \text{ lm each} = 8550 \text{ lm}$$

3. Determine the light loss factor (LLF):

$$LLF = 0.79 \text{ (standard value for typical commercial office space)}$$

4. Determine the coefficient of utilization (Cu):

$$Cu = 0.67 \text{ (typical value—consult manufacturer's catalog for more exact numbers)}$$

5. Insert these values into the equation:

$$\text{Number of lighting fixtures required} = \frac{Fc \times A}{FL \times LLF \times Cu}$$

$$= \frac{30 \times 420.75}{8550 \times 0.79 \times 0.67}$$

$$= 2.79$$

6. Round up to the next whole number to determine the total number of fixtures required.

Answer: With the provided specifications and fixtures, this room requires three lighting fixtures.

■ Determining Lighting Fixture Location

The distribution of the luminance from lighting fixtures in an area is affected by the spacing between the fixtures and the mounting heights of the fixtures. **FIGURE 4-3A** illustrates the correct lighting fixture quantity and spacing criteria for a space requiring a luminace level of 30 fc. The greater the space between fixtures and the higher the mounting height from the surface to be illuminated, the less light that reaches the work surface as shown in **FIGURE 4-3B**. However, when fixtures are mounted too close together or at low ceiling heights, they cannot provide even light distribution throughout the area. Focused light levels may be desired in locations such as retail areas where the goal is to spotlight certain objects, but in general office environments, even light distribution is required.

Designers can use the **spacing criteria** data that lighting fixture manufacturers provide in their specification sheets to calculate the recommended minimum spacing between lighting fixtures based on mounting height. These data are provided as two multipliers that can be used to calculate the recommended maximum mounting distance for the lighting fixtures:

- Row multiplier: The row multiplier is used to determine the mounting distance of the lighting fixtures in the end-to-end direction, sometimes referred to as the Angle 0 measurement.
- Column multiplier: The column multiplier is used to calculate the center-to-center distance across each fixture, sometimes referred to as the Angle 90 dimension.

These multipliers are applied to the mounting height of the lighting fixture from the surface to be illuminated. Once the mounting height of the lighting fixture and the distance to the illuminated object are known, the designer multiplies the distance to the illuminated object from the lighting fixture (in feet) by the row and column spacing criteria. The number generated is the recommended mounting space between the lighting fixtures. When designers adhere to manufacturer-recommended spacing criteria, they can design the lighting system to provide the most even luminance levels throughout rooms.

■ Creating the Lighting Plan

To design a lighting plan, the designer requires floor plans or **reflected ceiling plans** from the architect. These plans illustrate the ceiling types such as acoustical tile type ceiling or drywall ceiling for the space to be designed. This information is necessary for the designer to choose the proper types of lighting fixtures and their mounting methods.

Light plans are often produced on a separate lighting plan sheet because they contain a large number of detailed elements that might be overlooked if included on the same pages as the power plan: Designing lighting and power plans on different sheets allows for better clarity. The detailed elements that are typically found on lighting plans are the following:

- Lighting fixture schedule
- Branch-circuit identification
- Control device (e.g., switch, sensor) type and location
- Raceway location
- Keynotes

For projects with typical T-bar grid and acoustical ceiling tile systems, the designer must obtain a plan illustrating the layout and design of the ceiling grid. From this plan, the lighting designer can then provide a design for the portion of the building to be constructed with a grid ceiling system to ceiling specifications. For areas of the facility where the ceiling is of a hard surface material such as plaster or wallboard, flush type lighting fixtures are typically specified. For any open ceiling areas with high ceiling heights such as manufacturing areas where the mounting height of the lighting fixtures is 15 ft or more, designers typically choose newer style, high-bay fluorescent or more traditional metal halide lighting fixtures. Both of these lighting fixture types are designed to be mounted at greater ceiling heights and to provide high luminance levels over large open areas.

Lighting Fixture Schedule

Designers create a **lighting fixture schedule** in the design plan to relay all the relevant lighting fixture details. Lighting fixture schedules list the following information for each fixture:

- Identification (alphabetical reference)
- Power rating in volt-amperes

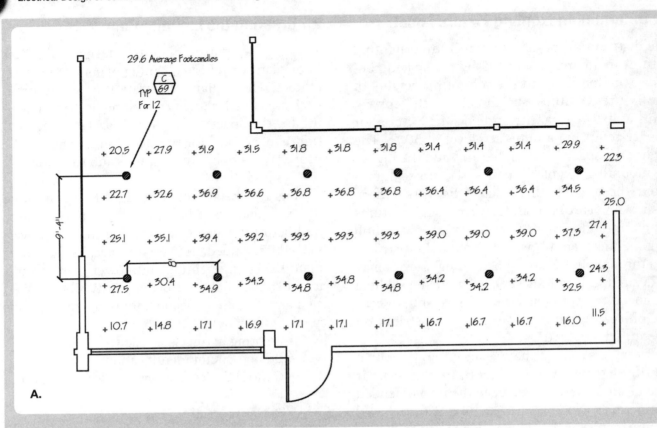

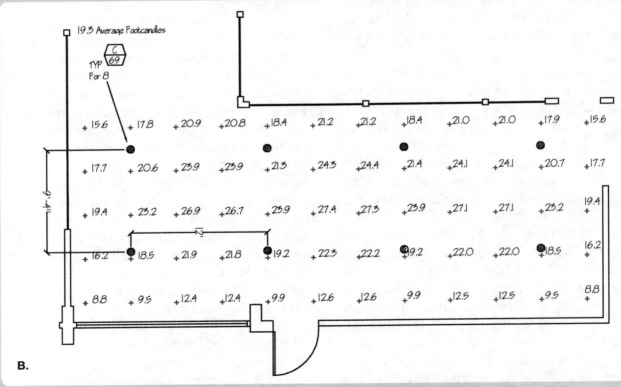

FIGURE 4-3 The distribution of light varies depending on the spacing between fixtures and their mounting heights.
A. A proper design has an average overall luminance level of 29.6 fc with 12 recessed lighting fixtures.
B. An improper design has only eight lighting fixtures and uses incorrect spacing criteria to achieve a luminance level of only 19.3 fc.

Determining Lighting Fixture Location

To determine lighting fixture location for fixtures with the following specifications:
- Lighting fixtures to be mounted in a ceiling 9 ft above the floor
- Work surface to be illuminated located 2.5 ft above the floor
- Lighting fixtures with a row multiplier of 1.25 and a column multiplier of 1.37

1. Determine the distance from the mounting height to the work surface:

$$9 \text{ ft} - 2.5 \text{ ft} = 6.5 \text{ ft}$$

2. Multiply this distance by the row and column multipliers:

$$\text{Row: } 6.5 \text{ ft} \times 1.25 = 8.13 \text{ ft}$$

$$\text{Column: } 6.5 \text{ ft} \times 1.37 = 8.91 \text{ ft}$$

Answer: The lighting fixtures should be mounted with an end-to-end dimension of approximately 8 ft and a side-to-side dimension of approximately 9 ft.

- Type of lighting fixture (e.g., flush mount, acoustical tile grid mount, high bay)
- Lamp type and quantity within the fixture
- Lighting fixture input watts
- Mounting method (e.g., recess mounted, surface mounted, pendant mounted, wall mounted)
- Brief description of the type of lighting fixture and variations
- Lighting fixture manufacturer name
- Manufacturer catalog numbers for lighting fixture types

Although there is no set industry format and most are created by designers, lighting fixture schedules must contain—at a minimum—the preceding information.

In the sample lighting fixture schedule shown in **FIGURE 4-4**, the first column, Fixture ID, lists the fixture alphabetical reference over the fixture's volt-ampere rating in a polygon symbol. The next four columns (under Fixture Type) list the type of fixture, such as incandescent, fluorescent, high-intensity discharge, or light-emitting diode (LED). Fixture Quantity is the total quantity of lighting fixtures (of the type specified). The next two columns (under Lamp) list the quantity of lamps per fixture and the wattage and type of the lamps. The next four columns (under Mounting) list the mounting method (recessed, surface, pendant, wall). The Description and Variations column lists any details that will be helpful to the installer during installation.

The final column, Manufacturer and Catalog Number, lists the manufacturer name and the catalog number of the lighting fixture.

The lighting fixture schedule lists each fixture using an alphabetical reference (e.g., A, B, C) that matches the reference at each lighting fixture on the design plan. No specifications designate which fixture receives which alphabetical reference. In general, designers assign each fixture a letter as they design it (the first fixture is labeled A, the second B, and so on). Designers might vary this system slightly when a specific lighting fixture from one manufacturer must be specified more than once because it includes additional components (such as a battery backup for emergency lighting). In this situation, designers usually identify the fixture without the additional components with a letter (e.g., A) and the fixture with additional components with the same letter and a number (e.g., A_1). When several lighting fixtures of the same type are located in the same general area, designers designate them as "TYP" (which stands for "typical") and indicate the quantity of that lighting fixture for that specific location. They insert this label on the plan adjacent to one representative lighting fixture to indicate that all lighting fixtures in that area are of the same type (see **FIGURE 4-5**). This method of identification helps to reduce the amount of information included on the print, which provides better overall clarity and makes the plan easier to read.

LIGHTING FIXTURE SCHEDULE

FIXTURE I.D.	FIXTURE TYPE				FIXTURE QTY	LAMP		MOUNTING				DESCRIPTION & VARIATIONS	MANUFACTURER & CATALOG NUMBER
	INCAND.	FLUOR.	H.I.D.	L.E.D.		NO.	WATTS AND TYPE	REC.	SURF.	PEND.	WALL		
A/88		X			28	3	F32T8	X				3 LAMP TROFFER	Lithonia 2SP8 3 32 A12 1/3 ADDE
B/36		X			84	1	F32TRT	X				FLUOR. DOWNLIGHT	Lithonia AF 1/32 8 AR MVOLT
C/69		X			76	2	F32TRT	X				FLUOR. DOWNLIGHT	Lithonia AF 2/32 8 AR MVOLT
D/185			X		14	1	150MH	X				METAL HALIDE DOWNLIGHT	Gotham Lighting AH 150M 8AR 277
E/368		X			37	6	F54T5 HO			X		HIGH BAY FLUOR.	Lithonia FSB 654L KDS 277 2/3
F/50	X				15	1	50W MR16			X		TRACK FIXTURE	Capri CV6117BK
G/112		X			5	4	F32T8			X		TANDEM INDUSTRIAL FLUOR.	Lithonia TL232MV
H/3.3				X	8		L.E.D.			X		EXIT LIGHT	Isolite CMB G U WW Z

FIGURE 4-4 A lighting fixture schedule lists important details about the selected lighting fixtures.

Branch-Circuit Identification

After the lighting fixtures for a particular project are designed, the designer must calculate the required number of **lighting branch circuits** necessary to serve the lighting system. Then, the designer can determine which branch circuit will serve each lighting fixture and identify the lighting branch circuit at each lighting fixture location. This iden-

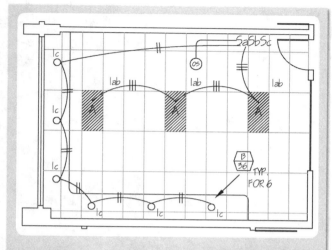

FIGURE 4-5 Identification method showing that all six lighting fixtures in this area are the same type.

tification provides a reference for the field installer, who needs to identify which lighting branch circuit is to serve the fixture.

Maximum Allowable Volt-Ampere Rating for Lighting Branch Circuits. The first step in determining the required number of lighting branch circuits is for the designer to calculate the maximum allowable volt-ampere rating for each of the lighting branch circuits that will serve the lighting system.

Most commercial office and manufacturing facilities are served by the electric utility with one of two types of 3-phase, 4-wire distribution systems: 120/208-volt systems or 277/480-volt systems. To calculate the maximum allowable volt-ampere ratings for lighting branch circuits, the designer must first determine the supply voltage. If a facility is served with a 120/208-volt, 3-phase, 4-wire system, the lighting system is served by a 120-volt system. If a facility is served with a 277/408-volt, 3-phase, 4-wire system, many times the lighting system is served with a 277-volt system even though within the facility **step-down transformers** will be installed to provide 120-volt supply for the general receptacle loads in offices (see Chapter 5).

When available, designers should choose the higher potential of 277 volts. This allows for larger quantities of lighting fixtures to be served by each branch-circuit because of their higher allowable volt-amperes and requires fewer electrical materials (branch-circuit wiring, raceways, etc.) to serve the lighting system. Designers must remember, however, that if any one lighting branch circuit that serves many lighting fixtures should fail, large areas of the facility may be without lighting until power is restored. Though seemingly contradictory, by limiting the area served by any one 277-volt lighting fixture branch circuit, designers can allow a greater number of lighting fixtures to be served by a lighting system operating on a lower voltage. As with branch circuits supplied to equipment, the *National Electrical Code* (NEC) requires that lighting branch circuits classified as continuous type loads be designed to operate at not more than 80 percent of their ampacity [210.19(A)].

Number of Required Lighting Branch Circuits. Once the serving voltage for the lighting system and the maximum allowable volt-ampere rating are determined, the designer can calculate the minimum quantity of required branch circuits for lighting loads. This provides the minimum number of branch circuits required, although additional lighting branch circuits can be designed to distribute the lighting circuits more evenly throughout the facility. In the design phase, as branch circuits are assigned to each lighting fixture, the designer must tabulate a volt-ampere count to ensure that any one lighting branch circuit does not serve more than its allowable volt-ampere rating.

Control Devices

All fixtures in a facility must have control devices to turn them on and off. To meet energy-saving requirements, lighting systems are now required to include specialized control methods such as **occupancy sensors** that automatically turn off lighting fixtures after a predetermined amount of time.

Occupancy sensors work by sensing motion in a room, and they turn off lights after a preset amount of time when they detect no motion, thereby

For each lighting fixture on a plan, the branch circuit serving the fixture must be identified adjacent to the fixture symbol.

Determining Maximum Volt-Ampere Ratings for Lighting Branch Circuits

To determine the maximum volt-amperage for a lighting branch circuit with the following specifications:
- 277-volt or 120-volt branch circuit
- 20-A branch-circuit rating

Use the following formula:

Supply voltage × Size of the branch-circuit overcurrent protective device (typically 20 A) × 80 percent

1. For a 277-volt lighting branch circuit:

$$277 \text{ V} \times 20 \text{ A} \times 0.80 = 4432 \text{ VA}$$

Answer: For this serving voltage, the maximum allowable volt-amperes that could be served for each 20-A lighting branch circuit is 4432 VA.

2. For a 120-volt lighting branch circuit:

$$120 \text{ V} \times 20 \text{ A} \times 0.80 = 1920 \text{ VA}$$

Answer: For this serving voltage, the maximum allowable volt-amperes that could be served for each 20-A lighting branch circuit is 1920 VA.

---

reducing energy consumption. Designers calculate the number of sensors required for a space based on the dimensions of the area and manufacturer specifications on the sensing range of the sensors.

For more advanced lighting designs, **programmable lighting controllers** can be installed to provide more sophisticated lighting control. These sensors allow the lighting system to be reprogrammed as the facility's needs change.

In the areas where programmable controllers are used, designers must include additional **bypass switches** so that personnel in the facility can access lighting when the program has shut down the system. With bypass switches, personnel can turn on lighting fixtures individually in the spaces they need to occupy without illuminating the entire facility, which reduces lighting energy consumption.

As per energy-saving requirements dictated by lighting energy codes (described later in this

---

Determining the Number of Required Lighting Branch Circuits

To determine the minimum number of required lighting branch circuits for a design with the following specifications:
- Total combined lighting volt-ampere rating of 12,650 VA (based on lighting fixture quantities and the sum their individual volt-ampere ratings)
- 277-volt, 20-A lighting branch circuit

1. Calculate the maximum ratings using the formula given previously:

$$277 \text{ V} \times 20 \text{ A} \times 0.80 = 4432 \text{ VA}$$

2. Divide the volt-ampere total for all lighting fixtures to be installed by the total allowable volt-ampere rating for the lighting branch circuit:

$$\frac{12{,}650 \text{ VA}}{4432 \text{ VA}} = 2.85 \text{ branch circuits}$$

Answer: A minimum of 3 lighting branch circuits is required.

chapter), the lighting fixtures in all spaces must be controlled by more than one switching method to allow for a 50 percent reduction in the lighting energy used. For example, a window can provide additional illumination during daylight hours, so occupants can turn off some lighting fixtures to achieve 50 percent reduction in lighting usage when daylight is sufficient. Designers can achieve the 50 percent reduction by providing two light switches to control alternate lighting fixtures or by providing a method to control the individual lamps in a lighting fixture with multiple lamps. For example, for a four-lamp fluorescent lighting fixture, one switch controls two of the lamps and a second switch controls the other two lamps. This is a very good method of providing 50 percent energy reduction that dims every fixture 50 percent while maintaining an average overall luminance level within the space. This type of switching is commonly referred to as the **A/B switching method**. The A/B switching method is typically used in conjunction with occupancy sensors.

When A/B switching methods are installed in conjunction with occupancy sensors, they meet all the mandated energy codes at very reasonable installation costs. Often, with this type of lighting control, two wall switches are located together at the entry door(s) to the space and the occupancy sensors are installed in a ceiling-mounted device. Alternatively, a wall switch and occupancy sensor may be manufactured as one device and mounted at the entry door, which eliminates the need for a separate ceiling-mounted occupancy sensor.

Lighting Branch-Circuit Raceways

The lighting design plan must illustrate which switching devices control which lighting fixtures and the location of the switching devices. Designers must illustrate all the control devices to the lighting fixtures and their respective raceways on the lighting plan. The lighting branch-circuit raceway layout should illustrate the following information:

- Control device locations
- Which lighting fixtures are controlled by which device
- Number of conductors within the raceway

Depending on the amount of detail designers choose to include in a lighting design plan, sometimes the plan illustrates only the routing of the raceways from the control device to the lighting fixtures being controlled. This is permissible, but at a minimum, designers must indicate the branch-circuit number serving each lighting fixture and the switching methods used for fixture control. For lighting design plans, as with power plans, the greater the detail illustrated, the higher the quality of the design, which helps reduce confusion and mistakes during the installation process. In addition to reducing delays and unnecessary costs, detailed plans provide greater safety during maintenance.

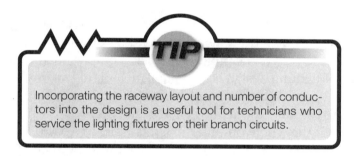

Incorporating the raceway layout and number of conductors into the design is a useful tool for technicians who service the lighting fixtures or their branch circuits.

■ Including Keynotes Sections

A high-quality design also includes a keynotes section. Keynotes sections provide additional specific information about lighting components and their applications, such as lighting fixture mounting heights and mechanical support methods. They also provide a reference to a more detailed wiring diagram that illustrates specific wiring information for advanced lighting controls. There are no specific requirements as to how keynotes should be structured because the method varies widely based on the scope of the project. Providing this additional information to assist installers greatly reduces the chance of installation errors and improper operation of devices and equipment.

Lighting designs must conform to energy code requirements. Designers who are not familiar with state or local energy code requirements can hire consultants to complete the necessary documentation.

Complying with Energy Code Requirements

In addition to the lighting needs of the facility, designers must consider nationally mandated lighting energy codes set forth by the U.S. Department of Energy. These energy codes dictate the amount of energy that lighting systems can use to ensure that all lighting fixtures and their components provide high quality and efficient lighting. Some states (such as California) have adopted additional standards that must be met for lighting design. Before beginning a lighting system design, designers should always obtain all required documents from the appropriate jurisdiction to ensure that they apply the mandated guidelines correctly. Designers adhere to national and state energy codes by designing lighting systems that do not exceed allowable energy levels based on watts per square foot (W/ft^2).

Interpreting Energy Code Design Methods

Designers can employ various methods to meet energy code requirements, depending on the details and needs of the project. The least complex method of lighting design is the **complete building method**. This method uses an allowable power density value to compute the lighting energy level evenly throughout the whole building. For example, in a 5500 ft^2 commercial building with an energy allowance of 1.1 W/ft^2, the lighting system cannot exceed 6050 W (5500 ft^2 × 1.1 W/ft^2). Energy allowances vary based on location and jurisdiction. **TABLE 4-3** lists typical lighting energy allowances for commercial buildings when designs incorporate the complete building method.

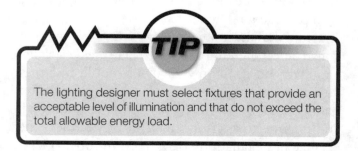

The lighting designer must select fixtures that provide an acceptable level of illumination and that do not exceed the total allowable energy load.

TABLE 4-3 Lighting Design Procedures: Complete Building Method

Primary Function	Allowable W/ft^2
General commercial and industrial work—High bay	1.1
General commercial and industrial work—Low bay	1.0
Industrial and commercial storage	0.7
General office work	1.1
Retail or wholesale	1.5*
Other	0.6

*For retail and wholesale stores, the complete building method may only be used when the sales area is 70% or greater of the building space.

The complete building method is easy to apply, but it does not allow for lighting power variance within a building that may be necessary for specialty lighting such as in retail showroom spaces. Office space illumination levels typically are not as high as those necessary for retail product displays. These different levels of illumination require different power levels based on watts per square foot. For example, a typical office space requires about 1.1 W/ft^2, whereas a retail showroom can require approximately 2.0 W/ft^2. When designers use the complete building method, they calculate the building illumination at an overall value of 1.1 W/ft^2, so the higher energy level needed in the showroom is not easily achieved. Additionally, as per lighting energy codes, retail or wholesale areas cannot comprise more than 70 percent of the total area of the building when the complete building method is used. Because of these restrictions, for applications with multifunctional areas, designers can use an alternative design method called the area category method.

The **area category method** identifies each space within the building by title and task and provides a total energy allowance for the building that can be divided based on the needs of each space. This method allows for variable lighting levels so that areas requiring more illumination (such as display areas) can exceed mandated energy allowances and areas where lighting may not be as crucial (such as stairways) can be designed below the mandated energy allowances. With this method, designers must be sure that the total energy used in the building lighting system does not exceed a calculated building maximum. To create a lighting design using the area category method, designers must first determine the total allowable energy level for each of the defined spaces, and then total the individual values to determine the total energy allowance for the building ensuring the total does not exceed the mandated maximum power values. **TABLE 4-4** lists typical allowable power density levels for this design method.

TABLE 4-4 Lighting Design Procedures: Area Category Method

Primary Function	Allowable W/ft²
Commercial and industrial storage	0.6
Corridors, restrooms, stairs, and support areas	0.6
Electrical, mechanical rooms	0.7
General commercial and industrial work	
High bay	1.1
Low bay	1.0
Precision	1.3
Commercial office space	1.2
Retail merchandise sales, wholesale showrooms	1.7
Waiting area	1.1
Other	0.6

When calculating total power density for a space, the lighting fixtures may be of more than one type with different input wattage levels as long as their combined wattages do not exceed the maximum energy allowance.

Energy Management

If a lighting design dictates that the mandated energy allowances must be exceeded, additional **lighting energy-saving devices** can be incorporated into the lighting system design. These devices can offer "credits" that reduce the energy levels to within mandated levels. For example, if a lighting design exceeds the mandated energy allowance by 1000 W, a credit of 1000 W from use of the additional energy-saving devices can be applied, which places the overall design within acceptable limits. Designers must consult the jurisdiction that dictates the lighting energy requirements to obtain information on which types of lighting energy-saving devices are approved for lighting energy credits.

Energy-saving devices typically include occupancy sensors, automatic shutoff devices such as time clocks, and programmable lighting control systems that automatically adjust lighting power levels. These devices can reduce a facility's overall energy consumption to comply with the maximum energy allowance.

Using the Area Category Method to Meet Energy Code Regulations

To use the area category method to calculate the maximum allowable lighting power in watts for a commercial building with the following specifications:
- General office space of 500 ft²
- Retail space of 1000 ft²

1. Determine the maximum allowable watts for the office space:

$$1.2 \text{ W/ft}^2 \times 500 \text{ ft}^2 = 600 \text{ W}$$

2. Determine the maximum allowable watts for the retail area:

$$1.7 \text{ W/ft}^2 \times 1000 \text{ ft}^2 = 1700 \text{ W}$$

3. Determine the total lighting power level for the building by adding these values together:

$$1750 \text{ W} + 600 \text{ W} = 2350 \text{ W}$$

Answer: For this application, a maximum of 2350 W can be used for the lighting system.

Note: If the retail area requires a higher luminance level for product display, additional lighting fixtures may be required, which will exceed the maximum level of 1.7 W/ft². This is permissible if the lighting in the office area can be redesigned using lighting fixture(s) that require lower input power levels and the total input power level for both spaces does not exceed the maximum power level of 2350 W. If this is the case, continue with the following steps.

4. Increase the lighting power level of the retail area to 2 W/ft²:

$$2 \text{ W/ft}^2 \times 1000 \text{ ft}^2 = 2000 \text{ W}$$

5. Readjust the maximum lighting power levels of the office area to accommodate the increased lighting power level of the retail area:

2350 total allotted watts − 2000 watts for retail area = 350 maximum watts allowable for the office space

6. If you can redesign the lighting for the office space at 350 W or less while maintaining the designed foot candle levels, and the overall design does not exceed 2350 W, then this application is permissible:

Retail space at 2000 W + Office space at 350 W = 2350 W

Wrap Up

■ Master Concepts

- Lighting systems include lighting fixtures, operational components, and control devices such as switches and sensors.
- To determine the lighting requirements of a facility, the designer must consider the tasks being performed in each area and the level of illumination required to complete these tasks adequately.
- When selecting lighting fixtures, designers must consider the fixture style, efficiency, lamp type, and mounting method in addition to the important lumens per watt ratio, which affects the lighting fixture's CRI rating.
- Designers should consider a number of variables—namely, desired luminance level, area of the space, fixture lumens, light loss factor, and coefficient of utilization—to determine the number of fixtures required.
- Designers must consider relevant spacing criteria to determine proper lighting fixture location.
- A lighting plan includes a number of detailed elements and therefore often requires a separate lighting plan sheet to list the lighting fixture schedule, branch-circuit identification information, control device type and location, raceway location, and keynotes.
- Designers must adhere to all *NEC* guidelines in addition to any applicable mandated energy requirements.
- The two common design methods used to ensure that the lighting plan meets energy code requirements are the complete building method and the area category method.
- Designers can use additional energy-saving devices to gain lighting energy credits if a lighting design exceeds the maximum lighting energy allowance.

■ Charged Terms

A/B switching method A dual switching method to control lighting that reduces the connected lighting load by at least 50 percent, maintains reasonably uniform illumination, and helps meet mandated lighting energy requirements.

area category method A lighting design method that designers can use to meet mandated lighting energy codes. In this method, the designer assigns a maximum allowable watts per square foot level to specifically define areas to provide adequate luminance for the primary function of the occupancy type. This method allows for maximum power levels in any one space to be exceeded when other spaces can be designed at lower allowed levels. The net result is that total power used does not exceed the calculated maximum allowed value. (*See also* complete building method.)

bypass switch A switch installed to override any automatic lighting shut-off device (e.g., time clock).

color rendering index (CRI) The ability of a lighting source to correctly represent an illuminated object in relation to natural daylight.

complete building method A lighting design method used to meet mandated lighting energy codes by calculating the maximum allowable lighting power for a facility based on a maximum allowable watts per square foot value; a basic design method not suitable for areas with specialty lighting. (*See also* area category method.)

foot candle (fc) A measurement of illumination intensity. One foot candle is the intensity of light on a surface 1 foot from a lighting source of 1 candlepower.

lighting branch circuit A branch circuit that serves only lighting.

lighting energy-saving devices Devices such as multiple switches, time clocks, and occupancy sensors that can achieve lighting energy savings.

lighting fixture schedule A document included with the lighting design plan that lists the specific information for the lighting fixtures in a facility.

lighting system Components such as branch circuits, switching devices, and energy-saving devices that are associated with lighting fixtures and their control.

lumens per watt (lm/W) The ratio of light output (lumens) to input power (watts).

luminance level The amount of light projected on a work surface.

occupancy sensor A device that detects the presence of personnel in a space by passive infrared or ultrasonic methods; when used in a lighting system, the sensor switches lighting fixtures on and off as occupants enter or exit the space to help save energy.

programmable lighting controllers Microprocessor-based lighting controllers that can be programmed; their use results in greater lighting energy savings.

reflected ceiling plan A plan that illustrates only the location of lighting fixtures and the ceiling type in which they are to be installed.

spacing criteria Mounting height ratios provided by the lighting fixture manufacturer used in calculations to determine proper mounting distances between lighting fixtures (also called row and column spacing criteria).

step-down transformer A transformer that delivers a different utilization voltage; typically used in commercial applications to lower a 480/277-volt service to 120/208 volts for office spaces.

■ Check Your Knowledge

1. What is the average luminance level for a general office space?
   - A. 10 fc
   - B. 30 fc
   - C. 60 fc
   - D. 100 fc

2. Use the manual calculation method to find the required number of lighting fixtures in a general office space with the following specifications:
   - Total area of 1350 ft^2
   - Desired luminance level of 35 fc
   - 3-lamp fluorescent lighting fixtures that emit 2850 lm per lamp and that have a coefficient of utilization of 0.86 and a light loss factor of 0.85

   How many fixtures are required?
   - A. 5 fixtures
   - B. 10 fixtures
   - C. 8 fixtures
   - D. 12 fixtures

3. Find the row and column spacing dimensions for a lighting fixture that has the following specifications: row and column criteria are row = 1.27 and column = 1.39; the fixture is designed to illuminate a surface 48 in. above the floor and is to be mounted on a 10-ft ceiling.
   A. R = 7.62 ft, C = 8.34 ft
   B. R = 9.67 ft, C = 10.85 ft
   C. R = 4.65 ft, C = 7.95 ft
   D. R = 4.0 ft, C = 8.0 ft

4. Which of the following data is NOT included in a lighting fixture schedule?
   A. Volt-ampere rating
   B. Type of lamp utilized in the lighting fixture
   C. Location
   D. Mounting method

5. A/B switching methods are used to:
   A. provide alternate switching methods.
   B. conserve energy.
   C. conform to energy code requirements.
   D. All of the above

6. What is the maximum allowable energy allowance for a 2500 ft^2 space designed as office space? Use the area category method (see Table 4-4).
   A. 1500 VA
   B. 1750 VA
   C. 2750 VA
   D. 3000 VA

7. Approximately how many branch circuits are required for a lighting load of 66,480 VA served by a 277-volt supply with 20-A single-pole circuit breaker?
   A. 10 branch circuits
   B. 15 branch circuits
   C. 25 branch circuits
   D. 12 branch circuits

8. Approximately how many lighting fixtures with a 108-VA rating could be installed on a 120-volt branch circuit supplied by a single-pole 20-A circuit breaker serving a continuous load?
   A. 25 lighting fixtures
   B. 15 lighting fixtures
   C. 18 lighting fixtures
   D. 30 lighting fixtures

9. Which of the following should be illustrated at each lighting fixture location on a lighting plan?
   A. Fixture type
   B. Branch-circuit number
   C. Switching method
   D. All of the above

10. The notation "TYP for 6" on a lighting design plan:
    A. indicates that six lighting fixtures in the surrounding area are all of the same type and manufacturer.
    B. provides for better clarity.
    C. simplifies the identification of lighting fixtures.
    D. All of the above

You Are the Designer

Apply the knowledge you have gained from this previous chapter to your own electrical design. In this section you will:
- Develop a lighting plan including the following elements:
  - Lighting fixture type, quantity, and location (optional)
  - Lighting fixture schedule
  - Branch-circuit identification
  - Control device location
  - Raceway location
  - Keynotes section
- Define applicable energy codes
- Use the area category method to design, and adjust if necessary, lighting power density levels to meet energy code requirements
- Incorporate energy management devices into the design
- Design a panelboard for the lighting system

■ About Your Project

To complete your task, you must know the following details about your project:
- Primary room functions (e.g., office, reception, etc.)
- Allowable W/ft^2 values (presented in Table 4-4) for each of the rooms in the facility

■ Resources

To develop this part of your design, you need the following resources from the *Student Resource CD-ROM*:
- Plan sheet E_2
- Blank lighting fixture schedule or partially completed lighting fixture schedule
- Excel spreadsheet "W/ft^2"
- Electrical symbol list

■ Get to Work

If time permits, you can design a lighting system in its entirety using a blank floor plan and select all the lighting fixture types.

If time does not permit this amount of research, you can choose to use the partially completed lighting design where the lighting fixture types, quantities, and spacing criteria have been determined. For this option, you must determine the volt-ampere rating for each of the lighting fixtures based on manufacturer and catalog information. For example, for Fixture C in the Open Office Area, you must obtain the volt-ampere rating for a recessed downlight that incorporates a two-lamp, 32-W, TRT type lamp. (If you choose this option, you should skip to the section titled "Creating the Lighting Plan" later in this chapter.)

Selecting Lighting Fixtures

If you are creating your own lighting design, you must first determine which type of lighting fixtures you will use. It is recommended that you use recessed mounted fluorescent type lighting fixtures in all office, open, storage, and similar areas. Incorporate traditional recessed mounted 2 ft by 4 ft grid-mounted fluorescent fixtures with two or three fluorescent 32-W type T8 lamps and 6-in. to 8-in. recessed mounted fluorescent lighting fixtures with compact fluorescent lamps in your design. Newer high-bay fluorescent fixtures or more traditional high-bay lighting fixtures (e.g., metal halide) are more appropriate for the manufacturing area.

Begin by gathering catalogs or searching online to select lighting fixtures and gather the following information from the manufacturer data sheets:
- Lighting fixture type (e.g., fluorescent, high-intensity discharge [HID], LED)
- Catalog number
- Spacing criteria
- Coefficient of utilization (if available)
- Light loss factor (if available)
- Lumens per lamp (supplied by lamp manufacturer)
- Volt-ampere rating

Note that the light loss factor value is automatically calculated if you use design software; for the purpose of this exercise for manual calculations, you can use a typical value of 0.79.

Add the required information to the lighting fixture schedule (provided on the *Student Resource CD-ROM*) and keep hard copies of all manufacturer data in the Lighting section of your project binder. Note where you obtained the information in case you need to find any additional information about the lighting fixture.

Calculating the Number of Fixtures Required

Once you have selected lighting fixtures, you can calculate the number of lighting fixtures required for each space. In addition to the data outlined earlier, you need to determine the desired luminance level for the space (see Table 4-1). For this project, the office space requires a luminance level of approximately 30 fc, and the manufacturing area requires approximately 50 fc. Use the following sample Calculation Example to guide you in determining the required number of lighting fixtures for each area. Enter the designed power levels for each space into the *Watts per Square Foot Table* on the *Student Resource CD-ROM*.

Lighting design software may also be used to help calculate the number of lighting fixtures required. For example, **FIGURE 4-6** shows how the luminance level was calculated using lighting design software for Office 9 with a minimum of 20 fc and a maximum of 42.2 fc, with an average of 29.2 fc, which meets the average of 30.0 fc designed.

When designing lighting for any space where the total lighting volt-amperes designed is less than the maximum allowed, the difference can be applied to other areas if needed, as long as the overall design does not exceed the total allowance for the entire facility.

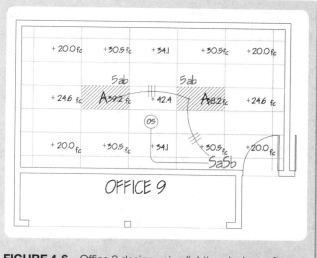

FIGURE 4-6 Office 9 design using lighting design software.

Calculating the Number of Lighting Fixtures Required for Office 9

To calculate the number of fixtures required for Office 9:

1. Determine the desired luminance level in foot candles:
   The desired luminance level is 30 fc (provided earlier).
   $$fc = 30$$

2. Determine the area in square feet:
   The area is 264 ft² (provided in *Watts per Square Foot Table* on the *Student Resource CD-ROM*).
   $$A = 264$$

3. Calculate the fixture lumens:
   This office uses fluorescent recessed fixtures: three T8 lamps with 2850 lm per lamp.
   $$3 \times 2850 \text{ lm} = 8550 \text{ lm}$$
   $$FL = 8550$$

4. Determine the light loss factor:
   For this project, the suggested guideline of 0.79 is used.
   $$LLF = 0.79$$

5. Determine the coefficient of utilization:
   This value was obtained from the manufacturer's specification sheet.
   $$Cu = 0.67$$

6. Use this data to calculate the correct number of fixtures.

$$\text{Number of lighting fixtures required} = \frac{fc \times A}{FL \times LLF \times Cu}$$

$$= \frac{30 \times 264}{8550 \times 0.79 \times 0.67}$$

$$= 1.75$$

(round up to the next whole number)

(continues)

Calculating the Number of Lighting Fixtures Required for Office 9 (*continued*)

Answer: Two three-lamp lighting fixtures must be installed to achieve the desired luminance level of 30 fc for Office 9.

Next, you must ensure that this number meets all applicable lighting energy requirements.

7. Determine the energy requirements for this space:

   Reference Table 4-4 to find that this office has a 1.2 W/ft² energy allowance.

8. Multiply the energy allowance by the area to determine the maximum allowable watts:

$$1.2 \text{ W/ft}^2 \times 264 \text{ ft}^2 = 317 \text{ W}$$

9. Calculate the total energy levels of the lighting fixtures selected for this space:

   Number of fixtures × Watts per fixture = Total lighting power level

   2 fixtures × 88 W (obtained from manufacturer specification sheets) = 176 W

10. Confirm that the total lighting power level is less than the maximum allowable watts for this space:

$$176 \text{ W} < 317 \text{ W}$$

Answer: The chosen fixtures will not exceed the maximum allowable energy requirements for this space.

Note: When using the area category design method, the difference between the maximum and consumed energy values (317 W − 176 W = 144 W) may be applied to another area of the design.

Determining Lighting Fixture Location

The next step is to determine lighting fixture location using spacing criteria. Gather the row and column spacing criteria from the manufacturer for each lighting fixture. For this project, assume that all ceiling heights in the general office, corridors, and restrooms are 9 ft and that the ceiling in the showroom area is 10 ft. Determine the distance from the lighting fixture to the surface to be illuminated. For this project, the counter spaces are approximately 2.5 ft off the finished floor.

To determine the mounting distances between lighting fixtures, multiply the distance to the working surface by the row spacing criterion to determine the minimum spacing of lighting fixtures end to end. Next, multiply the distance to the working surface by the column spacing criterion to determine the minimum spacing between the lighting fixtures. After you determine the location for the lighting fixtures for each area, illustrate the fixtures on your lighting plan using the appropriate symbol and letter as used in the fixture schedule (see Completed Plan Sheet E_2).

The actual mounting distances will vary based on the conditions of each space. For example, the acoustical tile ceiling in Office 9 has standard fixed dimensions of 2 ft by 4 ft. Therefore, the available end-to-end dimensions are 4 ft center to center and 2 ft across (see Figure 4-3). For this application, based on the direction of the installed ceiling grid and the calculated row and column data, the lighting fixtures in Office 9 must be installed

Determining Lighting Fixture Location for Office 9

To determine lighting fixture location for Office 9:

1. Determine the distance from the mounting height to the work surface by subtracting the distance from work surface to the floor from the ceiling height:

$$9 \text{ ft} - 2.5 \text{ ft} = 6.5 \text{ ft}$$

2. Multiply this distance by the row and column multipliers obtained from the manufacturer for the lighting fixture:

Row multiplier = 1.3

$$6.5 \text{ ft} \times 1.3 = 8.45 \text{ ft}$$

Column multiplier = 1.4

$$6.5 \text{ ft} \times 1.4 = 9.1 \text{ ft}$$

Answer: The lighting fixtures should be mounted with an end-to-end dimension of approximately 8.5 ft and a side-to-side dimension of approximately 9 ft.

---

with a row spacing of 8 ft center to center. Because this office requires only two lighting fixtures, the column spacing of any lighting fixtures was not required. If calculations determined that two additional lighting fixtures were required, then the lighting fixture spacing across would have been 8 ft as well. The next available spacing across would be 10 ft center to center, which exceeds the maximum recommended value.

Lighting for the Showroom. The lighting design for showrooms is more specific than it is for the general office areas. Unlike general office spaces that require an even, average luminance level throughout, showrooms require a higher luminance level directly on the product on display. This level is generally three times the luminance level of general spaces. For example, if a retail space has a general lighting luminance level of 20 fc, the product on display requires 60 fc. Usually, you can accomplish this by using recessed lighting for general lighting and using an additional light source such as track lighting for the product display (see *Lighting Fixture Schedule* for an example).

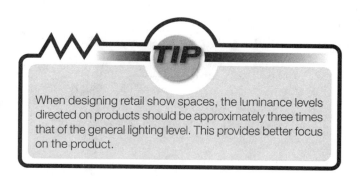

TIP

When designing retail show spaces, the luminance levels directed on products should be approximately three times that of the general lighting level. This provides better focus on the product.

For this project, it is recommended that you use a metal halide recessed fixture for the general lighting and a track lighting system with 12-volt MR16 lamps for product display. Both fixtures provide excellent color rendering index (CRI) values to illuminate the product with true color.

TIP

Metal halide and MR16 halogen lighting provides for much better color rendering for products.

Track lighting is calculated differently because of all the design considerations such as floor- or wall-mounted displays. Track lighting fixtures that use low-voltage lamps such as the MR16 have incorporated into the lighting fixture a step-down transformer that reduces the supply voltage, such as 120 and 277 volts, to the lower 12-volt potential, so additional components to lower the voltage are not necessary. Provide one track lighting fixture for every 3 linear feet of track. For display of wall-mounted products, track lighting should be installed approximately 3 ft from the wall (see Completed Plan Sheet E_2).

For the general lighting, select a recessed metal halide lighting fixture with lamp wattages between 50 W and 100 W. Determine the fixture quantities and locations using the examples in the previous section. Both general lighting and feature lighting for these types of spaces have several control switches, so design the general lighting for an overall luminance level of 30 fc for the entire room using the *Watts per Square Foot Table* provided. To calculate watts per square foot for the showroom, add the wattage values for all lighting fixtures installed (see "Determining Lighting Fixtures for the Showroom").

Determining Lighting Fixtures for the Showroom

To determine lighting fixtures for the showroom:

1. Determine the required number of lighting fixtures:
   - Desired luminance level (fc) = 30
   - Area = 1796 ft²
   - Lumen output for 100-Q metal halide lamp (FL) = 8700 lm
   - Light loss factor (LLF) = 0.65
   - Coefficient of utilization (Cu) = 0.59

$$\text{Quantity of fixtures} = \frac{fc \times A}{FL \times LLF \times Cu}$$

$$= \frac{30 \times 1796}{8700 \times 0.65 \times 0.59}$$

$$= 16.14$$

Answer: The showroom will require 16 to 17 lighting fixtures to achieve the desired luminance level.

Next, determine the lighting fixture locations.

2. Determine the distance from the mounting height to the work surface by subtracting the distance from work surface to floor from the ceiling height:

$$9 \text{ ft} - 2.5 \text{ ft} = 6.5 \text{ ft}$$

3. Multiply this distance by the row and column multipliers obtained from the manufacturer for the lighting fixture:

Row multiplier = 0.9

$$6.5 \text{ ft} \times 0.9 = 5.85 \text{ ft}$$

Column multiplier = 0.9

$$6.5 \text{ ft} \times 0.9 = 5.85 \text{ ft}$$

Answer: The lighting fixtures should be mounted with an end-to-end dimension of approximately 5.8 ft and a side-to-side dimension of approximately 5.8 ft, modified for the dimensions of the space.

Finally, check that the fixtures meet all applicable energy requirements.

4. Determine the energy requirements for this space:

Reference Table 4-4 to find that the showroom has a 1.7 W/ft² energy allowance.

5. Multiply the energy allowance by the area to determine the maximum allowable watts:

$$1.7 \text{ W/ft}^2 \times 1796 \text{ ft}^2 = 3053.2 \text{ W}$$

6. Calculate the total energy levels of the lighting fixtures selected for this space:

Number of fixtures × Watts per fixture = Total lighting power level

Fourteen 150-W metal halide recessed fixtures: 14 fixtures × 185 W = 2590 W

Fifteen 50-W track lighting fixtures: 15 fixtures × 50 W = 750 W

$$2590 \text{ W} + 750 \text{ W} = 3340 \text{ W}$$

7. Confirm that the total lighting power level is less than the maximum allowable watts for this space:

$$3340 \text{ W} > 3053.2 \text{ W}$$

Answer: For this application, the total wattage of the fixtures exceeds the maximum allowable energy requirements for this space. This design is acceptable using the area category method as long as the overall design for the facility does not exceed the total wattage allowed for the entire building.

Lighting for the Manufacturing Area. The manufacturing area lighting is similar to the general office space lighting except for the type of lighting fixtures chosen and the greater mounting heights of the lighting fixtures. Lighting fixtures in manufacturing areas are mounted at greater distances from the finished floor to allow the light to be broadcast over a larger area and avoid any interference with the manufacturing process.

In the past, metal halide (HID) fixtures were used, but many modern applications are designed with fluorescent lighting fixtures that include several lamps. These lighting fixtures have up to six 54-W T5 high-output lamps in each fixture with a total lumen output of 26,400 lm. These fluorescent fixtures are more energy efficient than metal halide high-bay lighting fixtures are; however, a larger quantity of fluorescent fixtures is required to meet the same luminance level. Less energy is used over the life of the lighting fixture, but initial installation costs are higher with fluorescent fixtures. Illustrate each selected fixture on the lighting plan using the appropriate lighting fixture symbol, and identify each lighting fixture using the proper letter in the fixture schedule (see Completed Plan Sheet E_2).

> **TIP**
>
> Even though high-bay fluorescent lighting fixtures are more efficient and will utilize less energy over their lifetime than metal halide lighting fixtures, the quantity of fixtures must be increased to achieve the same luminance level. Therefore, when selecting lighting fixtures for these areas, the cost and benefits should be evaluated for both types of lighting.

Determining Lighting Fixtures for the Manufacturing Area

To determine lighting fixtures for the Manufacturing Area 1 using fluorescent high-bay lighting fixtures with six 54-W T5 lamps:

1. Determine the required number of lighting fixtures:
   - Desired luminance level (fc) = 50 fc
   - Area = 6474 ft²
   - Lumen output (FL) = 30,000 lm (6 lamps with 5000 lm each)
   - Light loss factor (LLF) = 0.85
   - Coefficient of utilization (Cu) = 0.62

$$\text{Quantity of fixtures} = \frac{fc \times A}{FL \times LLF \times Cu}$$

$$= \frac{50 \times 6{,}474}{30{,}000 \times 0.85 \times 0.62}$$

$$= 20.47$$

Answer: The Manufacturing Area 1 requires 21 lighting fixtures to achieve the desired luminance level.

Next, determine the lighting fixture locations.

2. Determine the distance from the mounting height to the work surface by subtracting the distance from work surface to floor from the ceiling height:

   20 ft (as per keynotes section on the plan) − 2.5 ft = 17.5 ft

3. Multiply this distance by the row and column multipliers for the lighting fixture obtained from the manufacturer:

Row multiplier = 1.27

$$17.5 \text{ ft} \times 1.27 = 22.25 \text{ ft}$$

Column multiplier = 1.27

$$17.5 \text{ ft} \times 1.27 = 22.25 \text{ ft}$$

Answer: The lighting fixtures should be mounted with an end-to-end dimension of approximately 22.25 ft and a side-to-side dimension of approximately 22.25 ft, modified for the dimensions of the space.

Finally, check that the fixtures meet all applicable energy requirements.

4. Determine the energy requirements for this space:

   Reference Table 4-4 to find that this area has a 1.3 W/ft² energy allowance.

5. Multiply the energy allowance by the area to determine the maximum allowable watts:

$$1.3 \text{ W/ft}^2 \times 6474 \text{ ft}^2 = 8416.2 \text{ W maximum}$$

6. Calculate the total energy levels of the lighting fixtures selected for this space:

   Number of fixtures × Watts per fixture = Total lighting power level

   21 fixtures × 368 W (obtained from manufacturer specification sheets) = 7728 W

7. Confirm that the total lighting power level is less than the maximum allowable watts or this space:

$$7728 \text{ W} < 8416.2 \text{ W}$$

Answer: The chosen fixture design does not exceed the maximum allowable energy requirements of 8416.2 W for this space. The difference between the energy requirements (8416.2 W − 7728 W = 688.2 W) may be applied as a credit to other areas of the design if needed. When you have completed calculations for all designed areas, recalculate the total facility wattage to verify that the entire design is in compliance. As long as the overall design is under the total building allowable energy limits, the design is in conformance.

Lighting for the Assembly Area. In the completed lighting design plan, the lighting fixtures chosen for the assembly area are four-lamp, 8-ft fluorescent lighting fixtures that are designed to be suspended by chain. They are mounted at 12 ft above the finished floor.

TIP

More focused lighting is generally required for assembly areas (and areas for other related tasks). To accomplish this, bring the light source closer to the work surface.

Creating the Lighting Plan

Whether you have selected lighting fixtures yourself or used the partially completed plan, you should now have all necessary information in the Lighting section of your project binder, including a lighting fixture schedule listing lighting fixture volt-ampere values. The next step is to determine the required number of branch circuits that will serve the design.

It is recommended that you use a 277-volt system. In the power plan, a 480-volt, 3-phase, 4-wire panelboard is used for equipment located in the manufacturing area. This panelboard can be used to serve both the office areas and manufacturing area lighting systems (see panelboard P_1 in the completed plan). If you choose to develop a new panelboard that will serve only lighting loads, the number of panelboards is left to your discretion. Use the following steps:

1. Start by calculating the combined volt-ampere rating for all lighting fixtures installed in the facility by referencing the lighting fixture schedule.
2. Multiply the volt-ampere ratings for each fixture by the number of fixtures installed and add these values together to determine the total volt-ampere rating for all fixtures. (If you develop separate panelboards for the office and manufacturing areas, keep these totals separate.)
3. Divide the total volt-amperes by the maximum allowable volt-amperes for the voltage of the branch circuits (illustrated on completed P_1 on E_3) to determine the number of required branch circuits.

Determining the Number of Branch Circuits

To determine the minimum number of branch circuits for a project with the following specifications:
- Combined lighting of 28,306 VA
- 277-volt, 20-A branch circuits

1. Calculate the maximum volt-ampere rating:

$$277 \text{ V} \times 20 \text{ A} \times 0.80 = 4432 \text{ VA}$$

2. Determine the number of branch circuits:

$$\text{Number of branch circuits} = \frac{\text{Total lighting load}}{\text{Maximum volt-ampere rating}}$$

$$= \frac{28{,}306}{4{,}432}$$

$$= 6.3 \text{ (rounded up to the next whole number)}$$

$$= 7$$

Answer: A minimum of seven lighting branch circuits is required.

Note: This is just the minimum requirement. In completed panel schedule P_1, a total of nine lighting branch circuits is used to separate the lighting branch circuits that serve the office areas from those that serve the manufacturing area. Keeping the lighting branch circuits that serve these areas separate is a good design consideration.

For this application, the quantity of lighting branch circuits would need to be greatly increased if a 120-V lighting system was to be used instead of a 277-V branch circuit. For a 120-volt, 20-A branch circuit:

$$120\text{-volt} \times 20\text{-A} \times 0.80 = 1920 \text{ VA maximum}$$

<u>28,306 VA required</u>

$$1920 \text{ VA} = 14.74$$

Answer: For this application a minimum of 15 lighting branch circuits would be required (instead of just 7 branch circuits for 277-V lighting branch circuits).

Creating the Panel Schedule

The next step is to assign each branch circuit a number, as it will be distributed from the panelboard, using the template provided on the *Student Resource CD-ROM* for the appropriate volt-ampere values. For example, a 277-volt lighting system needs a 3-phase, 480-volt panel schedule, and a 120-volt lighting system needs a 3-phase, 208-volt panel schedule. If your lighting design uses the same 277/480-volt panelboard you designed for the manufacturing equipment, then you can continue with that panel schedule for this section. Use the following steps:

1. Enter the volt-ampere ratings for branch circuits you assign and assign branch-circuit numbers in the panel schedule so that they achieve a balanced distribution.
2. Under the OUTLETS/LTG column, enter the quantity of lighting fixtures served by each branch circuit.
3. Verify this by referencing the individual phase totals at the bottom of the panel schedule.
4. If necessary, make appropriate corrections by reassigning branch circuits to alternate branch-circuit numbers.
5. After you complete the branch-circuit distribution in the panel schedule, reference the appropriate branch circuit that will serve each lighting fixture at each lighting fixture on your design (see Completed Plan Sheet E_2 for an example).

Control Devices. The next step in completing the lighting design is to determine where the switches and other control devices will be located. It is recommended that for simplicity in this design you use toggle switches and occupancy sensors. By using these two types of devices together, you can meet the mandated energy requirements for control of lighting systems.

Many of the switches will be located near the entry door to an office space or corridor. For other open areas, where placement may not be as obvious, a good design method is to provide switches in areas where personnel would enter and proceed through the facility. For example, the completed lighting plan (see E_2) includes switches at the main entry, waiting area, and Office 8. As personnel enter open area 1, switches are provided to illuminate lighting in the space, and additional lighting switches are installed as personnel might proceed through the facility.

To meet lighting energy requirements, each space must have switching methods that provide for a 50 percent reduction in lighting. This is noted on the completed lighting plan at each switch location with the designation "A/B." For example, the switch in the waiting area is marked "Sab," and lighting fixtures are labeled "1a" and "1b." This signifies that the two switches control alternate lighting fixtures that allow

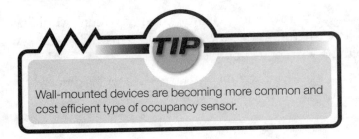

TIP
Wall-mounted devices are becoming more common and cost efficient type of occupancy sensor.

for a 50 percent reduction. Office 14 is marked "Sab" at the switches and "5ab" at the lighting fixture, which designates that each lighting fixture has multiple controls for the lamps within the lighting fixture. Manufacturers often provide the ability for groups of lamps within one fixture to be controlled separately. For fluorescent lighting fixtures that have two or fewer lamps, this is not possible, and the switching must be done by alternating the controlled lighting fixtures as was done in the entry area. Regardless of the method used, provisions must allow for the 50 percent reduction except in corridors, restrooms, and areas of less than 100 ft^2 (such as the storage and equipment rooms).

Regardless of the square footage, all areas are required to have occupancy sensors as an additional form of lighting control. These devices are illustrated on the plan with the designation "OS." If ceiling-mounted occupancy sensors are used, then the proper symbol is illustrated on the plan. If an occupancy sensor is incorporated into the switch, then use the "OS" designation at the switch location.

Lighting Branch-Circuit Raceways. With all the lighting fixtures and their switching methods complete, the next step is to illustrate the raceways from each of the lighting fixtures to the appropriate switches and indicate the number of conductors within each raceway. For example, the lighting in the waiting room illustrates the use of alternate lighting fixture switching methods and raceways with three conductors (switch leg A, switch leg B, and one grounded conductor); note that the last lighting fixture has only two conductors. Office 14 illustrates the use of controls for multiple lamps within a fixture. Both examples assume the power from the branch circuit is installed at the switch location and the only conductors necessary from the raceway at the switch to the lighting fixtures are the appropriate switch legs and the grounded conductor.

Reviewing the Lighting Plan

Finally, review your lighting plan to ensure that it contains all the following information:
- All lighting fixtures
- Lighting fixture identification (as listed in the lighting fixture schedule) at each fixture location
- Switching for all lighting fixtures with proper identification
- Branch-circuit identification at each fixture location
- Occupancy sensor controls

CHAPTER 5

Distribution Systems

Chapter Outline

- Introduction
- Single-Line Diagrams
- Designing Electrical Distribution Equipment
- Selecting and Sizing Distribution System Components
- System Grounding
- Using Conductors in Parallel

Learning Objectives

- Create a single-line diagram to industry standards.
- Calculate the required amperage sizes for the distribution equipment necessary for an electrical design.
- Generalize system requirements and criteria into a single-line diagram.
- Interpret the grounding requirements for an electrical system and apply them into an electrical design.

Electrical Design of Commercial and Industrial Buildings

■ Introduction

All facilities have a main point of entry for the electrical service that originates at the serving utility and terminates in the main electrical switchboard cabinet(s). From this main switchboard, all power is distributed to other panelboards serving lighting and power branch circuits.

Switchboards

When a facility requires more than 200 A, the distribution system will be a switchboard system. A **switchboard** is an electrical cabinet or cabinets (depending on the electrical requirements) that has provisions for the service entrance method, utility metering, and overcurrent protective devices that serve distributions to both the panelboards and larger branch circuits that serve larger motor loads (see **FIGURE 5-1**). Switchboards are larger than the typical wall-mounted electrical services that serve distributions of 200 A or less. These cabinets typically range in size from 30 in. to 36 in. in width to 90 in. tall and house **service entrance conductors**, **utility metering equipment**, the **main disconnecting means**, overcurrent devices for distribution to panelboards, and larger branch circuits.

■ Single-Line Diagrams

All the power requirements for all components such as the general branch circuits, specialized equipment branch circuits, and lighting system are brought together to develop a single-line diagram of the distribution system. A **single-line diagram** (also called a one-line diagram) is a simple diagram that illustrates all the information and requirements of the main electrical distribution system (see **FIGURE 5-2**), including the following:

- Service entrance equipment
- Distributions
- Branch-circuit distributions
- Distribution transformers
- System grounding methods

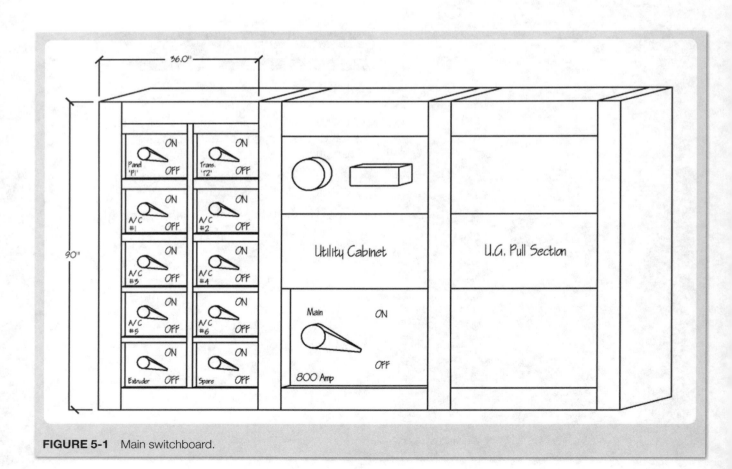

FIGURE 5-1 Main switchboard.

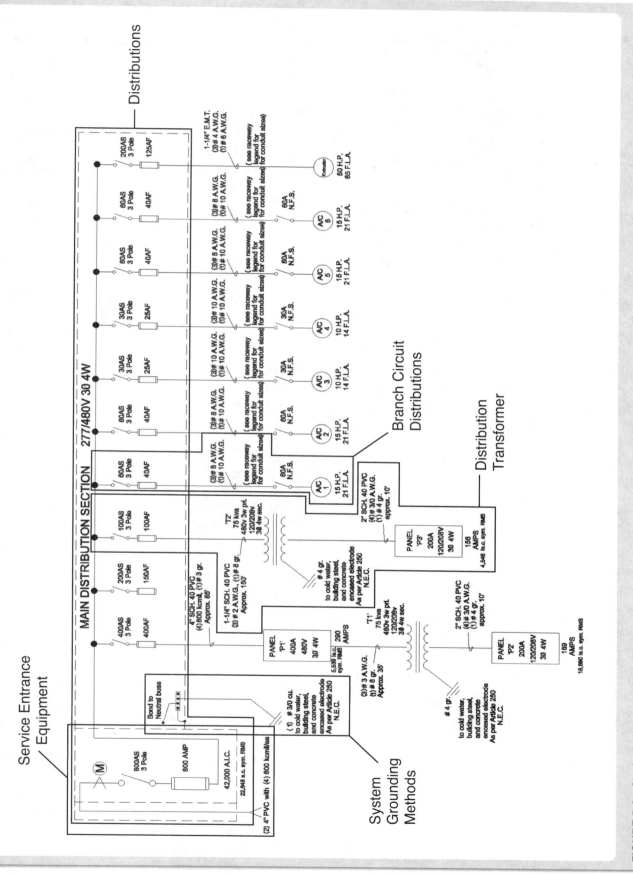

FIGURE 5-2 Single-line diagram. (Note: this diagram is also included in the *Student Resource CD-ROM*.)

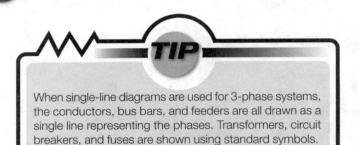

TIP: When single-line diagrams are used for 3-phase systems, the conductors, bus bars, and feeders are all drawn as a single line representing the phases. Transformers, circuit breakers, and fuses are shown using standard symbols.

The simple format of the single-line diagram clearly conveys all the required information about the main service entrance and distribution sections that serve the panelboards, transformers, and other large loads within the facility. Single-line diagrams do not provide information for smaller branch circuits; this information is conveyed through panel schedules and raceway legends. Branch-circuit information is only included in the single-line diagram if the equipment is served directly from the main switchboard (as is the case with some large motors or machinery loads that are not derived from branch-circuit panelboards).

Single-line diagrams are not usually drawn to scale. Instead, they are drawn to an appropriate size based on the scope of the project so that they can include all the necessary information within a single plan sheet.

In a single-line diagram, the switchboard is represented as a dotted outline showing the number of electrical cabinets required. Switchboards are typically illustrated with utility metering requirements on the left and distributions such as panelboards on the right, though this is not necessarily the order in which they will be installed. Cabinetry is manufactured as sections, and the electrical designer determines the order of these sections based on the location of the switchboard within the facility and the requirements dictated by the serving utility for the point of entry. The associated electrical equipment, including utility metering equipment and disconnecting means, is illustrated in the single-line diagram as a large rectangle adjacent to the underground section with fuse or circuit breakers marked using the appropriate symbols.

Service Conductors

Service entrance conductors, as defined by the *National Electrical Code (NEC)*, are the conductors from the service entry point to the service main disconnecting means. These conductors are installed from the main switchboard to a connection point dictated by the serving utility, and, depending on the serving utility requirements, they can be installed as part of an overhead-type installation or in underground conduits.

For underground installations, the electrical contractor typically provides and installs any required underground conduits to specifications provided by the serving utility. The serving utility supplies and installs the service entrance conductors.

In overhead (aboveground) installations, the designer determines the raceway and conductor sizes based on the requirements of *NEC* Article 230, and the electrical contractor provides and installs them.

Utilities may have a service planner who can work directly with electrical designers and contractors. The service planner may provide a detailed plan indicating raceway size, location, and burial depth requirements for underground conduit installations or point of building entry for overhead type installations.

Distribution

The distribution section of the single-line diagram provides information such as style of protection and rated amperage for individual distributions from the main switchboard to the panelboards, transformers, and, if designed, large motorized equipment loads throughout the facility.

The switchboard manufacturer builds into the system the required number of appropriately sized overcurrent protective devices. If the switchboard uses a fused-based system, the size of the disconnecting means, fuse sizes, and the number of poles must be indicated. If the switchboard uses a circuit breaker–based system, the size of the circuit breaker and the number of poles must be indicated.

TIP: Designers must always consult the utility provider and local municipality during the design phase to determine the applicable requirements.

For example, in the sample single-line diagram in Figure 5-2, the distribution to A/C 1 is referenced as 60AS/40ATDF, 3-pole. This designation indicates that this disconnecting means is to be supplied with a three-pole 60-A disconnect switch containing 40-A time delay fuses.

Panelboard, Transformer, and Branch-Circuit Distributions

The panelboard, transformer, and branch-circuit distribution section of the single-line diagram indicates the following information:
- Equipment being served
- Overcurrent protective device size and type
- Raceway quantity and type
- Conductor quantity, size, and insulation type
- Distance to the equipment
- Panelboard size (in amperage), phase, and voltage

Unless the owner of the facility has special identification requirements, panelboard identification is left to the discretion of the designer. Typically, power panelboards are designated P_1, P_2, and so forth; panelboards that supply only lighting loads are designated L_1, L_2, and so forth; and panelboards that supply both lighting and power are designated LP_1, LP_2, and so forth. If a switchboard will supply a separate cabinet designed as a motor control center, the designation is typically MCC.

Distribution Transformers

The distribution transformers section of the single-line diagram includes:
- Transformer title (e.g., T_1)
- Transformer size (in kilovolt-amperes [kVA])
- Transformer primary and secondary voltages
- Size of the transformer(s) primary and secondary protection (in amperes)

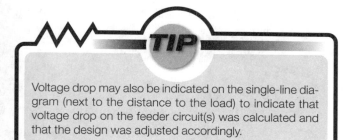

Voltage drop may also be indicated on the single-line diagram (next to the distance to the load) to indicate that voltage drop on the feeder circuit(s) was calculated and that the design was adjusted accordingly.

- Raceway size(s) for the primary and secondary sides of the transformer
- Conductor quantity and size for the primary and secondary sides of the transformer
- Distance of conductors serving the transformers
- System grounding electrode conductor size and grounding method

System Grounding Methods

The system grounding methods section of the single-line diagram indicates the following information:
- Size of the grounding electrode conductor(s) for the main switchboard
- Raceway method
- Bonding of the equipment

Designers should always confer with the local building department authorities, who may have specific regulations about which grounding methods are to be employed. If the jurisdiction is unknown at the time of the design, then the print should include a general note stating that the system grounding methods must conform to the requirements of the authority having jurisdiction. A note may also be added to the print stating, "Ground as per all applicable methods, based on *NEC* 250" to ensure that all applicable *NEC* requirements are met.

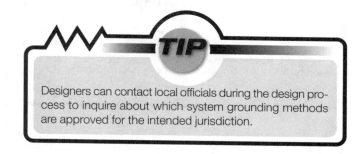

Designers can contact local officials during the design process to inquire about which system grounding methods are approved for the intended jurisdiction.

Designing Electrical Distribution Equipment

All distribution assemblies require a standard amount of physical space based on their voltage and amperage capacities. Electrical designers cannot arbitrarily choose a location for the main electrical distribution equipment. The serving utility may have location requirements, and any locations with electrical equipment must be designed with sufficient physical wall, ceiling, and floor space to meet the *NEC* working clearance requirements

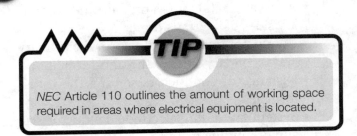

NEC Article 110 outlines the amount of working space required in areas where electrical equipment is located.

[110.26 and 110.30]. The switchboard manufacturer determines the amount of space required for the designed distributions and how many sections (cabinets) are required. Once this information is known, the electrical designer must ensure that the area where the cabinetry will be located is of sufficient size to meet the requirements of the NEC.

The width of the cabinets limits the quantity of distributions that can be derived from any one cabinet. If all the distributions are between 30 A and 100 A, then the switchboard could possibly be manufactured as a single cabinet. For distributions greater than 100 A, the switchboard may need to be manufactured with multiple cabinets because of the physical size of all distributions required.

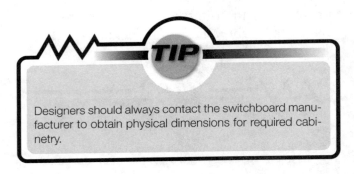

Designers should always contact the switchboard manufacturer to obtain physical dimensions for required cabinetry.

Selecting and Sizing Distribution System Components

The first step in the development of a single-line diagram is for the designer to make an overall determination of how many panelboards are required to serve all the branch circuits designed for general electrical needs, specialty equipment, and lighting. Panelboards are capable of serving approximately 42 branch circuits. Although large numbers of branch circuits can be served from any one panelboard, designs usually incorporate several panelboards. This is common practice because panelboards should be located within the approximate area where the loads are being served and because branch circuits serving general office areas and computers should be isolated from branch circuits that may serve motorized equipment loads. The starting and stopping of motorized equipment loads can sometimes cause momentary voltage surges, and if motorized equipment loads are supplied through the same panelboard as the one serving computer equipment and lighting circuits in office areas, these momentary voltage surges can cause problems with the computers or noticeable fluctuations in the lighting systems.

Sometimes facilities require multiple voltages, such as 480 and 208 volts for equipment and 120 volts for general office loads. The utility will not serve both voltage systems to such a facility. If, for example, the facility is served with a 480-volt system and it also requires a 120/208-volt system, the utility will serve the higher potential and the designer will have to provide for step-down transformers to obtain the additional voltage. In such situations, a panelboard cannot serve both 480 and 208 volts; therefore, additional panelboards are required to supply the lower voltage potential.

For most commercial applications, one panelboard is installed to serve the office area within a facility and one or more additional panelboards are installed in the manufacturing area based on the number of branch circuits required and their required operating voltages.

For a single-line diagram, the designer must determine the required number of panelboards based on the branch-circuit requirements, and then draw a basic outline of the distribution methods and the required equipment. For example, for a facility served with a 277/480-volt, 3-phase system and with the following requirements, a minimum of three panelboards should be designed:

- 480-volt, single- and/or 3-phase equipment loads
- 280-volt, single- and/or 3-phase equipment loads
- 120-volt general office loads

The panelboards would be assigned as follows:

1. One panelboard is located in the manufacturing area for the 480-volt equipment loads.
2. One panelboard is located in the manufacturing area for the 208-volt equipment loads.

3. One panelboard is located in the office area for the 120-volt general office loads.

Based on the minimum quantity of branch circuits required for each of the preceding applications, additional panelboards might be necessary.

With the number of panelboards and their branch circuits assigned, the designer can determine the amperage size of the panelboards and the conductor size of the feeders that supply them. Next, the designer creates a basic outline of the single-line diagram illustrating the distributions to the known panelboards. The designer will calculate the amperage of the panelboards and the size of the feeder conductors that serve them later in the design process.

Panelboards and Feeders

After a basic outline of the single-line diagram has been completed, the next step in designing the distribution system is for the designer to determine the appropriate feeder conductors and raceway sizes that will serve the panelboards. In the completed panel schedules, the total amperage listed at the bottom of the panel schedule can be referenced to standard National Electrical Manufacturers Association (NEMA) panelboard sizes (see **TABLE 5-1**). At a minimum, panelboards must have the amperage capacity to serve the designed loads determined by the panel schedules. Feeder conductors should then be sized based on the capacity of the panelboards.

TABLE 5-1 NEMA Standard Panelboard and Disconnect Sizes

Single-Phase Panelboards		Three-Phase Panelboards		Disconnect Switches	
40 A	200 A	60 A	225 A	30 A	800 A
70 A	225 A	125 A	300 A	60 A	1200 A
100 A	300 A	150 A	400 A	100 A	1400 A
125 A	400 A	200 A	600 A	200 A	1600 A
150 A	600 A			400 A	1800 A
				600 A	

As per the *NEC*, feeder conductors that serve panelboards are required to be rated at a capacity not less than the load to be served plus 125 percent for any part of the load that will be serving **continuous loads** [215.2(A)(1)]. If a panelboard will serve both noncontinuous and continuous loads (as is sometimes the case with general purpose receptacle and commercial lighting loads served by the same panelboard), then the panelboard and feeders must be sized to serve the noncontinuous load at 100 percent plus the continuous lighting load at 125 percent. The designer adds up all the loads and adds an additional 25 percent for any loads that have been determined to be continuous loads.

Distribution Transformers

When a transformer is installed to change one serving voltage to another to serve loads with different voltage requirements, the transformer must be sized appropriately to serve the associated loads. The size of the transformer is based on the total capacity of the panelboard being served, not the total calculated load determined by the panel schedule. **TABLE 5-2** lists standard NEMA manufactured sizes for 3-phase transformers.

As per the *NEC*, transformers must have overcurrent protection on the primary side [450.3(B)]. When transformers serve branch-circuit panelboards, they are required to have overcurrent protection on secondary side [408.36(B)]. Therefore, in commercial installations, both primary and secondary protection is required when transformers are used to reduce a source voltage of 480 volts to a lower 120/208 voltage to serve panelboards serving loads with lower voltage requirements.

■ System Grounding

The proper grounding of an alternating current system is of utmost importance to protect personnel and equipment from the dangers associated with electrical shock. The *NEC* is very detailed and specific about the associated components required for proper grounding and how the process is to be completed for all devices and equipment [Article 250].

For any system that is required to be grounded, an **effective grounding path** starts with the proper grounding of the electrical service equipment, which also includes the grounding of any transformers installed in a facility. When all the required grounding components are properly designed and installed to the designed specifications, a designer can feel confident that the system adheres to the requirements and that a high degree of safety is present in the electrical system.

Designing Panelboards for Lighting Systems

To design a panelboard that serves a combination of continuous lighting and noncontinuous loads with the following specifications:

- 277/480-volt, 3-phase, 4-wire panelboard
- 28,306-VA load contribution for the lighting branch circuits
- 205,310-VA load contribution for other noncontinuous loads

PANEL	P1'								VOLTAGE / PHASE:			277/480 3PH. 4 W.				
LOCATION:	Manufacturing								BUSS:			400A				
FLOOR:	First								MAIN BREAKER:			M.L.O.				
MOUNTING:	Surface															
	CT. #	OUTLETS			VOLT AMPS			BKR/ POLE	BKR/ POLE	VOLT AMPS			OUTLETS			CT. #
		LTG	REC	MISC	A	B	C			A	B	C	MISC	REC	LTG	
Lighting Office East	1.	70			3880			20/1	20/1	3340					29	2. Lighting Open/Shop
Lighting Showroom	3.	33				2277					1863				27	4. Lighting Office
Lighting Open 1	5.	53					2532					4048			11	6. Lighting Manuf.
Lighting Manuf.	7.	11			4048				40/3	5813			1			8. CNC Lathe 8
Lighting Manuf.	9.	8				3312					5813					10.
Lighting Manuf.	11.	17					3006					5813				12.
CNC Mill 1	13.			1	5813			40/3	60/3	9411			1			14. CNC Mill 2
	15.					5813					9411					16.
	17.						5813					9411				18.
CNC Mill 3	19.			1	5813			40/3	30/3	3875			1			20. CNC Lathe 9
	21.					5813					3875					22.
	23.						5813					3875				24.
75 Kva Transformer	25.			1	20787			100/3	40/3	5813			1			26. CNC lathe 6
Panel 'P2'	27.					20388					5813					28.
	29.						19643					5813				30.
CNC Lathe 5	31.			1	5813			40/3								32. Space
	33.					5813										34.
	35.						5813									36.
CNC Lathe 7	37.			1	5813			40/3								38.
	39.					5813										40.
	41.						5813									42.
	sub total				51967	49229	48433			28252	26775	28960	sub total			

Total A Phase = 80219 va
Total B Phase = 76004 va
Total C Phase = 77393 va
LCL (25%) = 7077 va

Total load   240693 va /   830   290 Amps

1. Calculate the required additional 25 percent for the long continuous load by multiplying the long continuous load served by the panelboard by 25 percent:

$$\text{Lighting load} = 28{,}306 \text{ VA}$$

$$28{,}306 \times 0.25 = 7077 \text{ VA}$$

2. Sum all loads plus the long continuous load volt-amperes:

Noncontinuous load volt-amperes	205,310 VA
Continuous load volt-amperes	28,306 VA
25% long continuous load factor	7077 VA
Total connected load	240,693 VA

3. Use Ohm's law to calculate the total connected load in amperes:

$$\text{Amperes 3-phase load (I)} = \frac{VA}{\text{Volts (E)} \times 1.73}$$

$$= \frac{240{,}693 \text{ VA}}{480 \text{ V} \times 1.73}$$

$$= 290 \text{ A}$$

4. Reference manufacturer catalogs to select the correct size panelboard:

The closest NEMA sized panelboards rated at 480 volts are 300 A and 400 A. For this application with a total calculated value of 290 A, a 300-A panelboard meets all the requirements. Therefore, this application can be served with the 300-A panelboard. However, the calculated value of 290 A for this application is close to the maximum allowable amperage for a 300-A panelboard; therefore, the next higher size of panelboard can be used so that the facility has future capacity.

5. Reference *NEC* Table 310.16 to properly size the feeder conductors that will serve the panelboard:

If a 300-A panelboard is used, then per Table 310.16, the minimum size thermoplastic high water-resistant nylon-coated (Type THWN) copper conductor rated at 300 A = 350 kcmil.

If a 400-A panelboard is used, then per Table 310.16, the minimum size Type THWN copper conductor rated at 400 amperes = 600 kcmil.

Answer: A panelboard sized at 300 A would be fed with 350 kcmil feeder conductors. A panelboard sized at 400 A would be fed with 600 kcmil feeder conductors.

TABLE 5-2 Three-Phase Transformers: Full Load Currents

KVA	208V	240V	480V	2400V	4160V	4800V	7200V	8320V	11,500V	12,000V	13,200V
4.5	12.5	10.8	5.41	1.08	0.63	0.54	0.36	0.31	0.23	0.22	0.20
7.5	20.8	18.0	9.02	1.80	1.04	0.90	0.60	0.52	0.38	0.36	0.33
9.0	25.0	21.7	10.8	2.17	1.25	1.08	0.72	0.62	0.45	0.43	0.39
10	27.8	24.1	12.0	2.41	1.39	1.20	0.80	0.69	0.50	0.48	0.44
15	41.6	36.1	18.0	3.61	2.08	1.80	1.2	1.04	0.75	0.72	0.66
22.5	62.5	54.1	27.1	5.41	3.12	2.71	1.8	1.56	1.13	1.08	0.98
25	69.4	60.1	30.1	6.01	3.47	3.01	2.0	1.73	1.26	1.20	1.09
30	83.3	72.2	36.1	7.22	4.16	3.61	2.4	2.08	1.5	1.44	1.31
37.5	104	90.2	45.1	9.02	5.20	4.51	3.0	2.60	1.88	1.80	1.64
45	125	108	54.1	10.8	6.25	5.41	3.6	3.12	2.26	2.16	1.97
50	139	120	60.1	12.0	6.94	6.01	4.0	3.47	2.51	2.40	2.19
75	208	180	90.2	18.0	10.4	9.02	6.0	5.21	3.77	3.61	3.28
100	278	241	120	24.1	13.9	12.0	8.0	6.94	5.02	4.81	4.37
112.5	312	271	135	27.1	15.6	13.5	9.0	7.81	5.65	5.42	4.92
150	416	361	180	36.1	20.8	18.0	12.0	10.4	7.53	7.22	6.56
200	555	481	241	48.1	27.8	24.1	16.0	13.9	10.0	9.62	8.76
225	625	541	271	54.1	31.2	27.1	18.0	15.6	11.3	10.8	9.84
300	833	722	361.2	72.2	41.6	36.1	24.0	20.8	15.1	14.4	13.1
450	1249	1083	541	108	62.5	54.1	36.0	31.2	22.6	21.6	19.7
500	1388	1203	601	120	69.3	60.1	40.0	34.7	25.1	24.1	21.3
600	1665	1443	722	144	83.3	72.2	48.0	41.6	30.1	28.9	26.2
750	2082	1804	902	180	104	90.2	60.0	52.0	37.7	36.1	32.8
1000	2776	2406	1203	241	139	120.0	80.2	69.5	50.2	48.1	43.7
1500	4164	3608	1804	361	208	180.0	120.3	104.2	75.4	72.2	65.9

Note: 3-phase transformer sizes in bold are standard values most readily available.

Calculating Amperage Size of Panelboard and Transformer

To calculate amperage size of a panelboard, transformer, and secondary feeder conductors for a system with the following specifications:
- Primary voltage of 480 volts
- Secondary 3-phase voltage of 120/208 volts
- Calculated secondary load of 165 A at 120/208-volt, 3-phase, 4-wire potential

1. Refer to *NEC* Table 310.16 to determine the amperage size of the panelboard sufficient for the load to be served:

   For a calculated load of 165 A, the minimum required size panelboard is 200 A.

2. Properly size the transformer to serve the load:

   For this application, the calculated load is determined to be 165 A, so a 200-A panelboard is used. The transformer must be rated to serve 200 A, and the secondary feeder conductors that serve the panelboard must be rated at 200 A.

3. Properly size the secondary feeder conductors:

   A 200-A panelboard must be served with feeder conductors capable of supporting the 200-A panelboard. As per Table 310.16, size 3/0, copper Type THWN secondary feeder conductors are required.

Answer: A 75-kVA transformer with a maximum secondary amperage of 208 A at 120/208 volts supporting a 200-A panelboard with size 3/0 Type THWN copper conductors are necessary for this application.

Grounding Distribution Transformers

When distribution transformers are installed to change a voltage to a higher or lower potential, the new potential then becomes a **separately derived system** and must conform to the system grounding requirements as dictated by the *NEC* [250.20]. In typical distribution transformers, the secondary voltages are provided through the principle of magnetism and the primary windings and secondary windings are not electrically connected. In a system where all grounding requirements are adhered to, but because the secondary side of a transformer derives its potential through magnetic principles and is no longer connected to the primary side of the transformer, any grounding that may have taken place on the primary side is lost. This new potential on the secondary side is classified as a separately derived system, and this new system must be grounded [250.30].

To ground a system properly, a continuous, nonspliced, **grounding electrode conductor** is installed between the electrical service equipment and the first **grounding electrode** (also called the primary ground) (see **FIGURE 5-3**). From the point of connection at the primary grounding electrode, an additional bonding jumper of equal circular mil area should be installed to the second grounding electrode (and third, if required). It must be understood that the primary

Sizing a Grounding Electrode Conductor

To size the grounding electrode conductor (using the 75 kVA transformer and 200-A panelboard):

1. Refer to *NEC* Table 250.66, which illustrates the required grounding electrode conductor sizes for alternating current (AC) systems based on the size of the largest ungrounded conductor serving the system.

2. Determine the secondary feeder conductor sizes. (The new 120/208-volt, 3-phase system is a separately derived system and all the grounding requirements must be based on secondary values.)

   For this example, size 3/0 Type THWN copper conductors are used as secondary feeder conductors.

Answer: *NEC* Table 250.66 requires that a size 4 copper grounding electrode conductor be used for a secondary feeder size of size 3/0 copper.

NEC TABLE 250.66 Grounding Electrode Conductor for Alternating-Current Systems

Size of Largest Ungrounded Service-Entrance Conductor or Equivalent Area for Parallel Conductors[a] (AWG/kcmil)		Size of Grounding Electrode Conductor (AWG/kcmil)	
Copper	Aluminum or Copper-Clad Aluminum	Copper	Aluminum or Copper-Clad Aluminum[b]
2 or smaller	1/0 or smaller	8	6
1 or 1/0	2/0 or 3/0	6	4
2/0 or 3/0	4/0 or 250	4	2
Over 3/0 through 350	Over 250 through 500	2	1/0
Over 350 through 600	Over 500 through 900	1/0	3/0
Over 600 through 1100	Over 900 through 1750	2/0	4/0
Over 1100	Over 1750	3/0	250

Notes:
1. Where multiple sets of service-entrance conductors are used as permitted in 230.40, Exception No. 2, the equivalent size of the largest service-entrance conductor shall be determined by the largest sum of the areas of the corresponding conductors of each set.
2. Where there are no service-entrance conductors, the grounding electrode conductor size shall be determined by the equivalent size of the largest service-entrance conductor required for the load to be served.
[a]This table also applies to the derived conductors of separately derived ac systems.
[b]See installation restrictions in 250.64(A).

Source: NEC® Handbook, NFPA, Quincy, MA, 2008, Table 250.66

grounding electrode conductor must not be spliced between the electrical service and the primary grounding electrode and must be attached to the grounding electrode using approved methods [250.64]. From the point of connection at the primary ground, a bonding jumper of the same American Wire Gauge (AWG) size as the grounding electrode conductor must be connected between the primary ground and the secondary ground. The bonding jumper is allowed to be a separate conductor from the primary grounding electrode conductor, but the grounding path is more consistent and reliable when both are installed as one complete conductor.

The size of the grounding electrode conductor is based on the size of the largest ungrounded service entrance conductor—or, in the case of a transformer, the secondary feeder conductors—used to serve the panelboard. These form a new separately derived system and therefore should be treated as service conductors. The size of the service grounding conductor is based on the sizes found in NEC Table 250.66.

Grounding Service Entrance Equipment

Service entrance equipment is grounded in the same manner as the distribution transformers are. As with transformers, an electrical service supplied through a utility transformer is classified as a separately derived system. The grounding electrode conductors for the main service are sized according to NEC Table 250.66 and, for the grounding electrode conductor for the main switchboard, the largest service entrance conductor. Grounding of service entrance equipment can differ somewhat from transformers if the utility serves the facility through an underground system. When a facility is served with an underground type distribution, the service entrance conductors are typically supplied and installed by the utility. Therefore, the designer may not know the actual size of the conductors. When this is the case, the designer selects the grounding electrode conductor based on the size of the largest service entrance conductor required for the main electrical service based on the size of the main overcurrent device.

■ Using Conductors in Parallel

When currents 400 A or greater are required by service equipment, panelboards, and other equipment, the conductors to serve them can be very large in kcmil size. This can be costly because of conductor and raceways size and difficulty of installation. Conductors may be connected in parallel (electrically

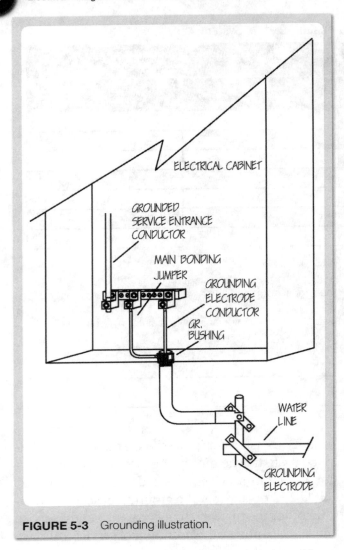

FIGURE 5-3 Grounding illustration.

For sizing grounding electrode conductors in parallel, the required size listed in *NEC* Table 250.66 is based on the equivalent circular mil area of the parallel service entrance conductors.

joined at both ends) to reduce the size of conductors [310.4]. This follows the basic Ohm's law principle for parallel circuits, which states that currents divide in parallel circuits. If each path has equal resistance and length, the currents divide equally. As per the *NEC*, conductors to be connected in parallel must be [310.4]:

- Sizes 1/0 and larger
- Equal in length and circular mil area
- Of the same material and insulation type
- Terminated in the same manner

Paralleling of conductors is not limited to two conductors per phase; when equipment requires large amounts of current, three or more conductors may be installed for each phase in parallel groups in multiple raceways as long as the minimum size of the paralleled conductors is greater than size 1/0 AWG/kcmil.

Paralleling Conductors

To parallel conductors for a system with a 400-A load with two parallel conductors:

1. Divide the total load in amperage by the number of conductors to determine the minimum amperage of each paralleled conductor:

$$\frac{400 \text{ A total load}}{2 \text{ paralleled conductors}} = 200 \text{ A each conductor}$$

2. Refer to *NEC* Table 310.16 to find that a 200-A copper Type THWN conductor should be size 3/0 AWG.
3. Ensure that the parallel conductor sizes are greater than size 1/0 AWG:
   A size 3/0 conductor is larger than a size 1/0 conductor.

Answer: For this application, two parallel size 3/0 conductors could be utilized as an option to using single 600-kcmil conductors. Please note that the paralleling of conductors is not limited to two, but can be three or more if the amperage of the load(s) is greater and the size of the paralleled conductors is not less than a size 1/0 conductor.

Wrap Up

■ Master Concepts

- Designers develop single-line diagrams to outline the electrical service entrance equipment and how the electrical system is distributed throughout a facility.
- Generally, transformers are required to provide for a change in voltage from the source voltage (at the service entrance point) for general and equipment electrical loads that operate at different voltages.
- Electrical service entrance equipment and transformers must be properly grounded to protect personnel and equipment.

■ Charged Terms

continuous load Any load where the maximum current is expected to continue for 3 hours or more [100].

effective grounding path A grounding path of low resistance that ensures, either through raceway methods or additional wiring methods, that the operation of protective devices will occur to isolate a faulted system and thus protect personnel from the dangers of electrical shock or explosion.

grounding electrode A conducting object through which a direct connection to earth is established [100].

grounding electrode conductor A conductor used to connect the system grounded conductor or the equipment to a grounding electrode or a point on the grounding electrode system [100].

main disconnecting means The main device that disconnects the supply conductors from all sources of supply.

separately derived system A premises wiring system whose power is derived from a source of electrical energy or equipment other than a service. Such systems have no direct electrical connection, including a solidly connected grounded circuit conductor, to supply conductors originating in another system [100].

service entrance conductors The conductors from the service entry point to the service main disconnecting means. These can be installed as part of an overhead-type installation or in underground conduits.

single-line diagram A simple diagram (also called one-line diagram) that illustrates all the information and requirements of an electrical distribution system.

switchboard An electrical cabinet, or cabinets (depending on the electrical requirements), that has provisions for the service entrance method, utility metering, and overcurrent protective devices that serve distributions to equipment.

utility metering equipment The components that make up the parts of an electrical cabinet used solely for the purposes of utility metering, such as kilowatt-hour/demand meters, meter testing points, and current transformers.

Check Your Knowledge

1. Which of the following is NOT illustrated on a single-line diagram?
   A. Size of the main overcurrent protective device
   B. Panelboard distributions
   C. Small equipment branch-circuit distributions
   D. Main switchboard grounding requirements

2. How many electrical switchboard cabinets are required for a project?
   A. One
   B. Two
   C. Three
   D. The number depends on project requirements.

3. For most typical underground service entrances, the minimum number of 10.2 cm (4-in.) raceway(s) required for every 400 A of electrical service is:
   A. one.
   B. two.
   C. three.
   D. four.

4. What is the minimum copper Type THWN primary conductor size required for a 112.5-kVA, 3-phase transformer with a primary voltage of 480 volts and a secondary voltage of 120/208 volts?
   A. Size 1 AWG
   B. Size 1/0 AWG
   C. Size 2/0 AWG
   D. Size 3/0 AWG

5. What is the minimum copper Type THWN secondary conductor size required for the transformer described in question 4?
   A. 250 kcmil
   B. 350 kcmil
   C. 400 kcmil
   D. 600 kcmil

6. What is the minimum required copper grounding conductor size for the 112.5-kVA transformer described in question 4?
   A. Size 1 AWG
   B. Size 1/0 AWG
   C. Size 2/0 AWG
   D. Size 3/0 AWG

7. What are the maximum allowable sizes for the primary and secondary overcurrent protective devices for the 112.5-kVA transformer described in question 4?
   A. 125-A primary, and 300-A secondary
   B. 150-A primary, and 350-A secondary
   C. 300-A primary, and 400-A secondary
   D. 250-A primary, and 600-A secondary

8. Feeder conductors supplied from the terminals of a transformer must be provided with overcurrent protection within the first:
   A. 5 ft.
   B. 10 ft.
   C. 15 ft.
   D. 50 ft.

9. Which of the following is NOT a standard NEMA 3-phase panelboard?
   A. 100 A
   B. 125 A
   C. 150 A
   D. 225 A

10. What is the required minimum copper grounding electrode conductor size for an electrical service fed with a parallel group of two 350-kcmil copper conductors for each of the phases?
    A. Size 2 AWG
    B. Size 1/0 AWG
    C. Size 2/0 AWG
    D. Size 3/0 AWG

11. A 200-A panelboard with a net computed load of 147 A must be fed with _____ Type THWN copper conductors.
    A. size 1/0 AWG
    B. size 2/0 AWG
    C. size 3/0 AWG
    D. 250-kcmil

You Are the Designer

Apply the knowledge you have gained from this previous chapter to your own electrical design. In this section you will:
- Create and draw a single-line diagram with the following components:
  - Switchboard
  - Service entrance conductors
  - Panelboard, transformer, and large branch-circuit distributions
  - System grounding
- Size the required components for the distribution system:
  - Panelboard and feeders
  - Distribution transformers
- Design a grounding system for the following components:
  - Transformers
  - Service entrance equipment
  - Conductors in parallel

■ About Your Project

To complete your task, you must know the following details about your project:
- Quantity of panelboards (previously determined)
- Which panelboards (1) will be served directly from the main switchboard and are supplied by 480 V or (2) will be served from distribution transformers to supply a 120/208-volt, 3-phase, 4-wire system
- Which large branch-circuit loads will be served directly from the main switchboard

■ Resources

To develop this part of your design, you need the following resources:
- Copies of all the panel schedules you have developed
- Single-line diagram on plan sheet E_3 on the *Student Resource CD-ROM*

■ Get to Work

In this part of the design, you bring together several components to develop the single-line diagram that forms the foundation for the overall designed load and the required size for the electrical service that will serve the facility.

Panel Schedules

Start by entering all required information into your panel schedules, beginning with the size of the overcurrent protective device that will serve each designed branch circuit. In commercial and industrial applications, general purpose receptacle branch circuits and many small specialty equipment loads (with volt-ampere ratings of 1800 VA or less) are typically served by single-pole 20-A circuit breakers. Use the designation 20/1 for the panel schedule to indicate that the branch circuit is served by a 20-A single-pole circuit breaker (see completed panel schedule P$_3$). For each panel schedule you have developed, determine which loads are to be served with these types of circuit breakers and enter the information in the panel schedule.

Next, calculate the size of the circuit breaker that will serve equipment loads served by panelboards and the size of overcurrent protection. Enter the amperes over the number of poles into the panel schedule (e.g., 80/3 for a 3-pole device). (See completed panel schedule P$_1$ for an example.) With the total calculated amperage at the bottom of your panel schedules, reference Table 5-1 to determine the minimum size panelboard required and enter that information in the panel schedule next to the heading Main Breaker.

Determining the Size of a Circuit Breaker Serving an Equipment Load

To determine the size of a circuit breaker serving an equipment load with the following specifications:
- 10–hp, 208-volt, 3-phase motor

1. Obtain the amperage of the motor based on voltage and horsepower. Use *NEC* Table 430.248 for single-phase motors and *NEC* Table 430.250 for 3-phase motors:

$$\text{Full-load amperes} = 30.8 \text{ A}$$

2. Multiply the full-load current of the motor by 250 percent (per *NEC* Table 430.52):

$$30.8 \text{ A} \times 2.50 = 77 \text{ A}$$

3. Reference *NEC* 240.6 for the standard size circuit breakers and size the overcurrent protection accordingly. Remember that if the value of the overcurrent device does not calculate to a standard value, you can use the next higher standard:

$$\text{Next higher standard overcurrent protective device} = 80 \text{ A}$$

Note: If the equipment to be served will be protected by a fused-based device, then a time delay type fuse should be used. For these applications, use the same procedure as described in this example, but multiply the full-load amperage by 175 percent, as allowed in *NEC* Table 430.52.

Distribution Transformers

When your design requires a distribution transformer to change the incoming 480 volts to a 120/208-volt, 3-phase, 4-wire system, you need to determine the size of the following components:
- Transformer (in kVA)
- Primary and secondary overcurrent protection devices
- Primary and secondary conductors
- Raceway types and sizes for the feeders for the primary and secondary conductors
- Grounding electrode conductor

As previously illustrated, the size of the transformer is determined by the amperage size of the panelboard it will serve. Start by identifying which panel schedules you have completed that will require a distribution transformer; these are all the panel schedules that operate at a 120/208-volt, 3-phase, 4-wire voltage.

To determine the required distribution transformer size, reference your panel schedule under the Main Breaker section. The value you entered here determines the minimum kilovolt-ampere size for the transformer that will serve the panelboard. For example, the completed panel schedule P_2 is protected by a 200-A main breaker; therefore, based on the transformer sizes in Table 5-2, a 75-kVA transformer is necessary.

kVA ratings for transformers must have the capacity to serve the rating of the panelboard they supply, even though the calculated loads in a panel schedule may be less.

When sizing the primary and secondary protection for transformers, you can use two methods based on *NEC* Table 450.3(B), which lists the maximum allowable primary and secondary protection sizes for transformers 600 volts or less. For a transformer with current greater than 9 A and that is protected on both the primary and secondary sides, the primary protection can be based on 250 percent of the full-load current and 125 percent of the full-load current of the device to be served (in this case, a panelboard). You use these maximum values typically only when a load that is to be served may have very high inrush currents that require the protections to be based on values above the actual rating of the device. For example, if the 200-A panelboard in the preceding example were to be protected at these higher values, then the primary and secondary protection sizes would be as illustrated in "Sizing Transformer Primary and Secondary Protections" below.

Sizing Transformer Primary and Secondary Protections

To size transformer primary and secondary protections with the following specifications:
- 75-kVA, 3-phase transformer
- 480-volt primary
- 208-volt secondary

1. Calculate the maximum value for the primary by multiplying the full-load amperes by 250 percent:

    90.2 A (as per Table 5-2) × 2.50 = 225.5 A

2. Reference *NEC* 240.6 to determine standard size protection:

    225 A

3. Calculate the maximum value for the secondary by multiplying the full-load amperes by 250 percent:

    208 A (as per Table 5-2) × 1.25 = 260 A

4. Reference *NEC* 240.6 to determine standard size protection:

300 A

Answer: If you use the maximum values permitted, this transformer could be protected on the primary side at 225 A and on the secondary side at 300 A.

Commercial and industrial applications typically do not require transformers to be protected at these maximum values because of the types of loads being served. To size protection for general use, distribution transformers are often protected at values based on the full load of the device to be served. For this application, the load to be served in a 200-A panelboard and the primary and secondary overcurrent protection sizes are determined. For example, for a 75-kVA, 3-phase transformer with a 480-volt primary and a 208-volt secondary, Table 5-2 shows that the full load at the primary is 90.2 A and at the secondary is 208 A, which—based on *NEC* 240.6 for a 200-A panelboard—would be sized at 200 A.

Transformer primary and secondary protection is typically provided based on a transformer's full load amperages. These values may be increased to the maximum value permitted by *NEC* Table 450.3(B) when the loads have significant inrush amperages. Large loads or loads with several motors require increased protection.

Because conductors are required to be sized to the load being served, the primary and secondary feeders for transformers are based on the full-load primary and secondary amperages. For example, for the same transformer, *NEC* Table 310.16 shows that Type THWN conductor for 90.2 full-load primary amperes would be a size 3 AWG conductor. For the 208 full-load secondary amperes it would be size 4/0 AWG. If this transformer was installed with an electrical metallic tubing (Type EMT) raceway, it would require three size 3 AWG feeders in a 2.5 cm (1-in.) Type EMT for the primary wire with four size 4/0 AWG Type THWN conductors for secondary feeders in a 5.1 cm (2-in.) Type EMT for a 120/208 system [Annex C, Table C.1]. The grounding electrode conductor is also sized based on the transformer secondary conductor sizes. As per *NEC* Table 250.66, the minimum copper grounding electrode conductor size would be size 2 AWG.

The grounding electrode conductor is based on the secondary feeder conductors for the new, separately derived system, not the primary conductor sizes.

Distribution System Raceways

Once you have determined and sized the panelboard and distribution transformers, the next step is to determine the proper raceway sizes for all distribution equipment. You have already located panelboards throughout the facility in previous phases of the design, so now you must design the raceway system that will serve them. Reference the completed power plan (E_1) and the completed distribution diagram (E_3). Note that transformer T_1 is located adjacent to the panelboard P_2, following the guideline that, when possible, it is best to locate the transformer as close as possible to the panelboard being served. Also note that the primary-side conductors are smaller (in AWG size) and carry lower amperage values than the secondary conductors. When panelboards are located in close proximity to the transformer serving them, the larger secondary conductors are much shorter in length than the primary conductors. This design helps with any voltage drop concerns caused by the length of a conductor and the amperages it must supply. Because primary currents are less, it is best to travel the distance to the panelboard with a primary conductor that serves less amperage and to supply the transformer secondary conductors with the higher amperages to the panelboard with a shorter-length conductor. Smaller conductors are also less expensive than larger conductors, so this design is more cost-effective (see **FIGURE 5-4**).

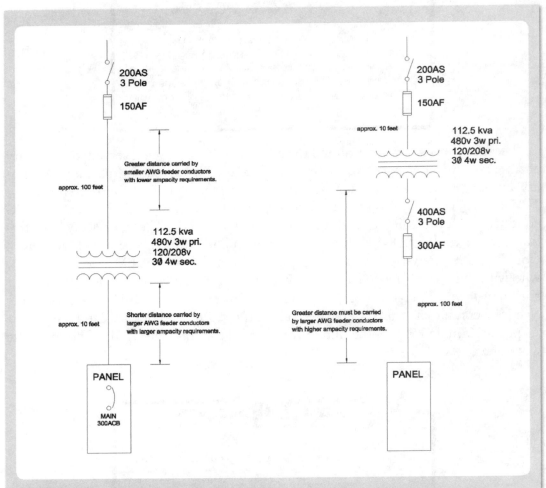

FIGURE 5-4 To help eliminate voltage drop and reduce installation costs, locate panelboards in close proximity to the distribution transformer serving them.

TIP

When a panelboard supplied by a transformer is fed from another panelboard, the load (in VA) must be included in the primary panelboard to properly reflect the load being by the primary panelboard (see completed panel schedule P_1).

The completed single-line diagram also shows that panelboard P_2 is supplied by panelboard P_1 through a 75-kVA transformer with maximum allowable primary of 90.2 A. The total volt-amperage values from panel schedule P_2 are included in panel schedule P_1 because panelboard P_1 serves both equipment loads and panelboard P_2. Before the addition of the 75-kVA transformer loads, it was determined that panelboard P_1 was also to serve the 75-kVA transformer and that the added load would not substantially increase the size of panelboard P_1; therefore, P_1 could also serve panelboard P_2. Panelboard P_2 could have been served directly from the main switchboard, but this would have made it necessary to run the primary feeder conductors for transformer T_1 from the main switchboard, resulting in greater lengths and a possible increase in the primary feeder conductor sizes (because of possible voltage drop). The disadvantage to serving panelboard P_2 through panelboard P_1 is that if panelboard P_1 is de-energized for repair or service, panelboard P_2 is also de-energized.

It is important to remember that when feeder conductors are connected to the secondary terminals of a transformer, protection must be provided in the first 10 ft [240.21(C)(2)]. This requirement is met when a panelboard is located within 10 ft of a transformer and the transformer has secondary protection (as is the case with panelboard P_2 having a main breaker). If the panelboard will be located more than 10 ft from the secondary of a transformer, an overcurrent device must be installed within the first 10 ft. **FIGURE 5-5** illustrates both options.

For your design, determine whether any panelboards served by transformers would be better served by supplying the panelboard through another panelboard. If you find one, revise your panel schedule file for the panelboard that will be serving a transformer and enter the volt-amperage values from the panelboard being served into the panel schedule for the panelboard serving the transformer (see completed E_3 for an example). If you alter your design, you must reevaluate the volt-amperage totals for the panelboard serving the transformer in case its size must be increased to accommodate the additional load. If so, you will also need to resize the main breaker, feeder conductors, and the raceway system serving the panelboard that serves the transformer.

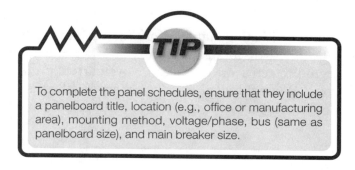

TIP

To complete the panel schedules, ensure that they include a panelboard title, location (e.g., office or manufacturing area), mounting method, voltage/phase, bus (same as panelboard size), and main breaker size.

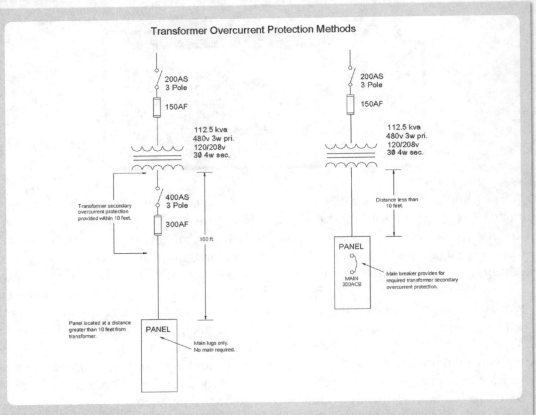

FIGURE 5-5 In the installation method on the left, the larger secondary feeder conductors carrying greater amperages are much longer in length, leading to voltage drop and greater installation costs. Overcurrent protection must be provided within 10 feet as per NEC 240.21(B)(1). In the illustration on the right, the primary feeder conductors carrying lower current amounts are carried the greater length, eliminating voltage drop concerns and helping lower installation costs. The main breaker in the panelboard also provides for the required transformer overcurrent to be within 10 feet of the transformer.

Equipment Loads Served Directly from the Main Switchboard

When equipment loads are directly served from a main switchboard, you must list the following information:
- Equipment title (e.g., A/C 1)
- Branch-circuit conductor sizes
- Equipment grounding conductor sizes (if applicable)
- Overcurrent protective device size and number of poles
- Additional disconnects (if applicable)
- Raceway types
- Horsepower and/or full-load amperage ratings
- Distances from main distribution system to loads

When installing branch circuits to air-conditioning units, additional disconnects are required at the air-conditioning unit locations for disconnection of power for servicing the units. When these disconnects are provided, they are not required to have additional overcurrent protective devices because overcurrent protection is provided at the source of supply. When installed, the devices are installed as nonfused devices and are illustrated on the single-line diagram as "nonfused disconnect" or "NFD."

Serving larger, motorized loads directly from the main service eliminates the need to increase the size or add additional panelboards to serve these loads.

Completing the Single-Line Diagram

At this point in your design, you can now complete all the required information about your distribution system. Use completed E_3 as a reference and illustrate the following information on your plan:

1. Panelboard distribution layout
2. Panelboard feeder sizes
3. Raceway methods
4. Transformers
    - Titles (e.g., T_1)
    - Size (in kVA)
    - Primary and secondary voltage
    - Primary and secondary feeder sizes
    - Grounding method
5. Loads (in amperage, located at bottom of panelboard)

Most plan errors occur in single-line diagrams. Review the single-line diagram to ensure that all necessary information is included and correct.

CHAPTER 6

Load and Short-Circuit Calculations

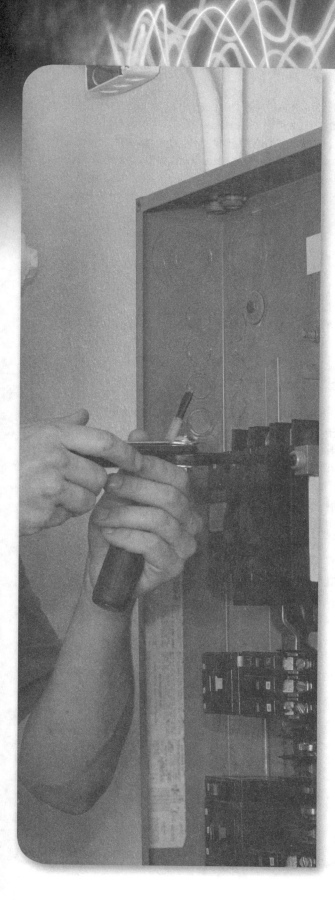

Chapter Outline

- Introduction
- Performing Load Calculations
- Performing Short-Circuit Calculations

Learning Objectives

- Interpret the importance of load and short-circuit calculations.
- Evaluate design calculations and revise load calculation values.
- Create a set of load calculations.
- Evaluate the procedure to complete a set of short-circuit calculations.
- Create a set of short-circuit calculations.
- Properly select equipment based on short-circuit calculations.

Introduction

Performing load calculations and short-circuit calculations is one of the final steps in completing an electrical design. At this stage in the design process, loads from all the various components of the design served by panelboards or the main switchboard must be brought together for load calculations that determine the amperage size of the main electrical service distribution equipment that will serve the electrical requirements of the facility.

If a short circuit or ground fault should occur anywhere in the electrical system, extremely high amperage values could flow through the system. If electrical components in the system are not properly protected from these currents, severe damage and possible electrical explosions can occur, destroying electrical equipment and causing injury or death to those working around the equipment (see **FIGURE 6-1**). Electrical designers perform short-circuit calculations to select protective devices to prevent any severe damage caused by short circuits or fault conditions.

When designers work with a service planner from the local utility, the load calculations will be provided to the planner, who works with the utility to design appropriate service entrance requirements. Accurate load calculations enable the serving utility to provide a service entrance plan that meets all the necessary requirements for approval.

FIGURE 6-1 Properly performing load and short-circuit calculations protects personnel and equipment from dangerous explosions caused by short-circuit or fault conditions.

Performing Load Calculations

It is not sufficient to simply add the panel schedule totals and equipment served directly by the main switchboard to calculate the total system electrical demands. As designers calculate electrical loads throughout the process and enter them into panel schedules, they enter these loads at 100 percent of rated capacity. **Load calculations** are necessary to evaluate what the total load will be for any one time during operation to properly size the panelboards and the feeder conductors that supply them. However, at 100 percent of rated capacity, these loads do not realistically represent the total electrical demand on the system at any one time because it is highly unlikely all loads will be operating at capacity at once. When calculating the actual demand for panelboards serving these types of loads, designers calculate allowances to more accurately reflect the demand on the electrical system.

For panelboards that serve equipment and motorized loads, allowances must be added to the total load. For the total calculated load from the panel schedule, designers must add an increase factor to ensure that the electrical system is designed to carry the equipment full-load amperages and can accommodate any additional amperage caused by inrush currents associated with motor loads.

Through the process of load calculations, designers evaluate these factors to determine the realistic electrical demand imposed on the electrical system at any one time.

General Electrical Loads

To determine the actual full load of general purpose receptacles, the designer must consider relevant demand factors. The *NEC* defines a **demand factor** as the ratio of the maximum demand to the total connected load under consideration [Article 100]. In

It is important to perform load calculations correctly. Underestimating a facility's demand limits its electrical capacity and ability to operate. Overestimating a facility's demand results in a larger than necessary distribution system with excessive equipment and labor costs.

Calculating Demand Factor (for General Purpose Receptacle Outlets)

To calculate demand factor for a commercial facility with the following specifications:
- General receptacle load of 25,000 VA

1. Calculate the first 10,000 VA at 100 percent:

$$10,000 \text{ VA}$$

2. Apply a demand factor of 50 percent to the remainder of load:

$$25,000 - 10,000 = 15,000 \text{ VA}$$

$$15,000 \times 0.50 = 7500 \text{ VA}$$

3. Add the totals to determine the total load after applying the allowable demand factors:

$$10,000 \text{ VA} + 7500 \text{ VA} = 17,500 \text{ VA}$$

Answer: The adjusted total for this load is 17,500 VA.

---

essence, demand factors are reduction allowances that can be applied to certain types of loads without high demand, such as the general purpose receptacle loads, to get a more realistic volt-amperage value. For example, the *NEC* permits general purpose receptacle loads to be computed at a value of 100 percent of the first 10 kVA of load and 50 percent of the remaining load [Table 220.44].

Specialized Electrical Loads

To size distribution systems for specialized equipment loads, the distribution equipment and conductors must be capable of supporting the higher inrush currents associated with motor loads without overload or voltage drop. Therefore, specialized equipment loads are calculated at 125 percent of the rated load of the device. When motor loads are supplied through panelboards, the 125 percent allowance is calculated for the largest motor supplied. If the allowance is added to the largest motor, then this allowance is sufficient for any motor(s) of smaller load supplied by the panelboard.

Demand factors that reduce the calculated load cannot be applied to specialized equipment loads or continuous loads such as lighting loads.

Lighting System Loads

Because of their long operating hours, lighting loads are considered continuous loads. When panelboards serve lighting loads, the designer must calculate the demand on the system at a value of 125 percent of the full load value [215.2(A)(1)].

Distribution System Loads

When distribution equipment directly serves motor loads without distribution through a panelboard, the designer must calculate the equipment demand at 125 percent of the highest-rated motor load plus the amperage for all the other motors being served.

Equipment should be sized based on the total load derived from the panel schedule. Demand factors are used only to determine a more accurate load that may be imposed on the main electrical service equipment and should not be applied when sizing branch-circuit panelboard or feeder conductors.

When panelboards or distribution equipment serve both specialized equipment and lighting loads, the calculation must add both 125 percent of the lighting load and 125 percent of the specialized equipment load.

Determining Load Calculations for Motor Loads

To determine the load calculations when more than one motor is being served:

1. Determine full-load amperage of the motor loads in volt-amperes using NEC Table 430.248 for single-phase motors or NEC Table 430.250 for 3-phase motors:
   a. 10-hp, 3-phase, 230-volt motor = 28 A
   b. 15-hp, 3-phase, 230-volt motor = 42 A
   c. 25-hp, 3-phase, 230-volt motor = 68 A
2. Use Ohm's law to calculate this value in volt-amperes:
   For a 3-phase motor: Voltage (E) × Amperage (I) × 1.73
   a. 230 V × 28 A × 1.73 = 11,141 VA
   b. 230 V × 42 A × 1.73 = 16,712 VA
   c. 230 V × 68 A × 1.73 = 27,057 VA
3. Calculate 125 percent of the volt-amperes of the highest-rated motor in the group:
   $$27{,}057 \text{ VA} \times 1.25 = 33{,}821 \text{ VA}$$
4. Add the volt-amperes of the other motors in the group at 100 percent:
   $$11{,}141 + 16{,}712 + 33{,}821 = 61{,}674 \text{ VA}$$

Answer: The total demand for this load is 61,674 VA.

Additional 125 percent demand values for subsequent motor loads are not necessary. The 125 percent allowance for the highest-rated motor provides for any other smaller motors served. In this type of application, when any motors in the group are of the same amperage rating, the designer must calculate 125 percent demand factor for the highest-rated motor in the group and then add the remaining amperages. If all motors are the same amperage, the designer must calculate 125 percent for any motor in the group and add the remaining amperages.

■ Performing Short-Circuit Calculations

If a short circuit or fault condition were to flow in a system, extremely high, damaging amperages would flow through the system unless proper protection was in place. These amperages are not just limited to the main entry point of the electrical system within a facility. If a short circuit or fault occurs at a panelboard located away from the main electrical service, the amperages flow through the entire path from the point of service entry to the location of the fault or short circuit, causing extreme electrical stress to any equipment in the path such as panelboards, the feeders that serve them, and any branch-circuit protective device in a panelboard.

Designers perform a set of **short-circuit calculations** to determine these amperage values, and then select components for the electrical system with rated values that ensure the equipment is properly protected. If extreme amperages occur, they are limited to values within the rated, short-circuit limits of the electrical equipment. According to the NEC:

> The overcurrent protective devices, the total impedance, the component short-circuit current ratings, and other characteristics of the circuit to be protected shall be selected and coordinated to permit the circuit-protective devices used to clear a fault to do

Distribution equipment follows a rule similar to feeder conductors. To calculate ampacity, take 125 percent of the highest rated motor in the group plus 100 percent of the remaining motor amperages [430.24].

so without extensive damage to the electrical components of the circuit [110.10].

Short-circuit amperages in a circuit are caused by the high levels of current available if a short circuit occurs. These short-circuit values may be provided by the serving utility or calculated manually to determine the maximum short-circuit amperage available. Like any electrical circuit, one factor that determines the amount of amperage is the impedance (AC resistance) in the circuit. However, with short-circuit amperages, higher and more extreme values are possible. For a service entrance point in a facility, assuming an infinite value available, the amount of short-circuit amperage is limited only to the impedance in the circuit from a serving utility transformer and the total impedance in the service entrance conductors.

Once the total available short-circuit amperage is determined, then the designer evaluates each point in the system (e.g., panelboards, equipment) to determine how much of the total short-circuit amperage is available at each point, based on the impedance in the circuit.

Transformer Impedance

All transformers 25 kVA and larger must include the __transformer impedance rating__ on the nameplate [450.11]. Impedance ratings are given in percentages. Typical ratings are approximately 3 percent to 6 percent, which means that if the terminals of a transformer were purposely shorted out (nearly zero impedance) and voltage was supplied to them, the transformer would reach its full-load amperage when the input voltage reaches 3 percent to 6 percent of the full voltage rating of the transformer. For example, if a transformer with a 3 percent impedance rating was rated at 208 volts, and the leads of the transformer shorted as a result of a line-to-line short circuit, the transformer would deliver its full-load current at approximately 6.24 volts (which is 3 percent of 208 volts). With the transformer already reaching full-load current value at 3 percent of rated voltage, more extreme currents would be delivered when the transformer reaches 100 percent of rated voltage.

When designing an electrical plan, designers must complete the short-circuit calculations for several points in the distribution path, including the following ones:
- Utility transformer (typically provided by the serving utility)
- Main service entrance point
- Distribution panelboards
- Motor circuits
- Motor control centers

Determining the Available Short-Circuit Current for a Transformer

To determine the available short-circuit current for a transformer with the following specifications:
- 225 kVA, 208-volt, 3-phase transformer
- Impedance rating of 3 percent

1. Determine the full-load current of the transformer using Ohm's law:

$$\text{Amperes } (I) = \frac{\text{kVA}}{\text{voltage } (E) \times 1.73} \times 1000$$

$$= \frac{225 \text{ kVA}}{208 \times 1.73} \times 1000$$

$$= 625.3 \text{ amperes}$$

2. Divide the full-load current by the impedance rating of the transformer:

$$\frac{625.3 \text{ A}}{0.03} = 20{,}842.7 \text{ A}$$

Answer: The available short-circuit current for this transformer is approximately 20,000 A. If a line-to-line short were to occur, this transformer has the capability of delivering more than 20,000 A.

The calculations to determine the short-circuit amperages must be illustrated on the electrical plans to verify that all the components of the distribution system are capable of withstanding any damaging short-circuit and/or fault current values that might occur. There are no set industry standards as to how the calculations must appear on a plan, but they should be clear and concise.

Designers can complete short-circuit calculations manually or with the assistance of specialized software. The method most commonly used for smaller or less detailed projects is the **point-to-point method** (see the following Calculation Example).

First, the designer must determine the available short-circuit amperage at the terminals of the utility transformer that will serve a facility, and then at the main service. The value of short-circuit amperage available at the main service will be less because of additional impedances based on service entrance conductor type, conductor length, and type of raceway. A conductor's impedance is chosen from a predetermined value known as the conductor's **C value** (see **TABLE 6-1**). This value is the reciprocal of the conductor's impedance to 1 ft of cable and varies based on the type of raceway used. With metallic raceways, high levels of amperage are somewhat depleted by **counter electromotive forces** (also called "back voltages" or CEMFs), which develop by the induced voltages within the metallic raceway. In nonmetallic type raceways, no induced CEMFs are developed, resulting in higher levels of available short-circuit amperages.

Once the value of short-circuit amperage has been determined for the main switchboard, the designer can determine the value of short-circuit amperage available at each component served by the main switchboard, such as panelboards. Again, these values decrease somewhat based on conductor types, lengths, and raceway types.

Additionally, it is important to note that if a short circuit occurs when motors are running, the motors can contribute a substantial amount of counter electromotive force back into the system. Therefore, if the main service supplies motor loads, the designer must calculate the motor contribution component by calculating the inductance to resistance ratio (X/R). Although the derivation of this factor is beyond the scope of this text, it is sufficient for most purposes to multiply the sum of all full-load amperages for the motors by a factor of 4.

As mentioned previously, short-circuit amperages will flow through all components in the path including fusible disconnects, panelboards, and the circuit breakers contained within them. Therefore, all fusible disconnects, panelboards, and circuit breakers are rated by their size in amperage and by the short-circuit amperes that can safely pass through the equipment without causing damage. For example, a standard 20-A single-pole 120-volt circuit breaker used in common panelboards is rated for a maximum allowable current of 20 A with a maximum short-circuit rating of 10,000 A. Circuit breakers are also available with short-circuit values of 22,000 A, 42,000 A, or 65,000 A (see **FIGURE 6-2**). Fuses can have maximum short-circuit amperage values of 200,000 A and more.

TABLE 6-1 *C* Values for Conductors

Copper AWG or kcmil	Three Single Conductors	
	Conduit Type	
	Steel	Nonmagnetic
	600 Volt	600 Volt
14	389	389
12	617	617
10	981	981
8	1557	1558
6	2425	2430
4	3806	3825
3	4760	4802
2	5906	6044
1	7292	7493
1/0	8924	9317
2/0	10755	11423
3/0	12843	13923
4/0	15082	16673
250 kcmil	16483	18593
300 kcmil	18176	20867
350 kcmil	19703	22736
400 kcmil	20565	24296
500 kcmil	22185	26706
600 kcmil	22965	28033
750 kcmil	24136	28303
1000 kcmil	25278	31790

Determining the Available Short-Circuit Current Using the Point-to-Point Method

To determine the available short-circuit current using the point-to-point method for a transformer with the following specifications (as illustrated on the completed sample plan):

- 750-kVA 3-phase transformer
- 3.5 percent impedance rating
- 480-volt secondary voltage
- Main service located 100 ft from utility transformer
- Main service supplied with two 600 kcmil copper conductors per phase in non-metallic PVC raceway

1. Determine the full-load secondary amperage for the transformer using Ohm's law:

$$\text{Amperage } (I) = \frac{\text{kVA}}{\text{Voltage } (E) \times 1.73} \times 1000$$

$$= \frac{750 \text{ kVA}}{480 \times 1.73} \times 1000$$

$$= 903.2 \text{ A}$$

2. Determine the short-circuit amperage available at the transformer:

$$\text{Short-circuit amperage} = \frac{\text{Transformer full-load amperage}}{\text{Impedance rating}}$$

$$= \frac{903.2 \text{ A}}{0.035}$$

$$= 25{,}805.7 \text{ A}$$

3. Determine the f factor, which is an impedance adjustment factor:

$$f = \frac{(1.73 \times \text{Distance to main service from utility transformer} \times \text{Short-circuit amperes})}{(\text{Number of conductors per phase} \times \text{Conductor C value} \times \text{System voltage})}$$

$$= \frac{(1.73 \times 100 \text{ ft} \times 25{,}805.7 \text{ A})}{(2 \times 28{,}033 \times 480)}$$

$$= 0.17$$

4. Determine the percentage of the initial short-circuit amperage that will be available at the main service:

$$\text{Percentage available} = \frac{1}{(1 + f \text{ factor})}$$

$$= \frac{1}{(1 + 0.17)}$$

$$= 0.85$$

5. Determine the short-circuit amperage that will be available at the main service:

$$\text{Short-circuit amperage} = \text{Maximum short-circuit amperage} \times \text{Percentage available}$$

$$= 25{,}805.7 \times 0.85$$

$$= 21{,}934.8 \text{ A}$$

(continues)

Determining the Available Short-Circuit Current Using the Point-to-Point Method (*continued*)

Answer: The available short-circuit amperage available at the terminals of the main service will be 21,934.8 A.

If this transformer will serve motor loads, continue with the following steps:

6. Calculate the motor contribution component. (For this example, assume the main switchboard is serving a motor contribution of 177 A and has an X/R factor of 4.)

$$\text{Motor contribution component} = \text{Sum of motor full-load currents} \times \text{X/R factor}$$

$$= 177\text{ A} \times 4$$

$$= 708\text{ A}$$

7. Calculate the total available short-circuit amperage:

$$\text{Total short-circuit amperage} = \text{Available short-circuit amperage} + \text{Motor contribution component}$$

$$= 21{,}934.8\text{ A} + 708\text{ A}$$

$$= 22{,}642.8\text{ A}$$

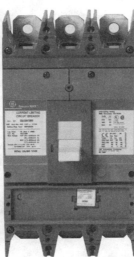

FIGURE 6-2 Circuit breakers are manufactured with various short-circuit ratings (e.g., 10,000 A; 22,000 A; 42,000 A; and 65,000 A).

Determining Available Short-Circuit Current for a Panelboard

To determine the available short-circuit current at panelboard P_1 with the following specifications:
- 480-volt, 3-phase panelboard
- Location 25.9 m (85 ft) from main switchboard
- 22,642.8 VA available short-circuit amperage at main switchboard
- Conductor C value of 28,033
- One 600 kcmil copper conductor per phase in non-metallic PVC conduit

1. Determine the f factor:

$$f = \frac{1.73 \times 85 \text{ ft} \times 22{,}642.8}{1 \times 28{,}033 \times 480} = 0.25$$

2. Determine the percentage of the initial short-circuit amperage that will be available at the main service:

$$= \frac{1}{(1 + 0.25)}$$

$$= 0.81$$

3. Determine the short-circuit amperage that will be available at the main service:

Short-circuit amperage = Maximum short-circuit amperage × Percentage available

$$= 22{,}642.8 \text{ A} \times 0.81$$

$$= 18{,}340.7 \text{ A}$$

4. Calculate the motor contribution component:

Motor contribution component = Sum of motor full-load current × X/R factor

$$= 153 \text{ A} \times 4$$

$$= 612 \text{ A}$$

5. Calculate the total available short-circuit amperage:

Total short-circuit amperage = Available short-circuit amperage + Motor contribution component

$$= 18{,}340.7 \text{ A} + 612 \text{ A}$$

$$= 18{,}952.7 \text{ A}$$

Answer: The total available short circuit amperage at panelboard P_1 is 18,952.7 A.

If short-circuit calculations determine that the short-circuit amperage values available in an electrical system could exceed the maximum short-circuit amperage rating of any electrical equipment or device in the path of a short circuit, the equipment must be rated at or above the calculated available short-circuit value. Alternatively, a greater degree of protection can be provided in the circuit ahead of the equipment to reduce or eliminate these damaging currents.

Circuit breakers do not stop or limit these extreme amperage values; their maximum short-circuit ratings only specify that these devices will safely pass the high amperage levels without damage to the device itself. Fuses are designed to handle short-circuit amperages without damage to the fuse and to block the flow of the short-circuit amperages at the fuse. If a short-circuit calculation determines that a short-circuit amperage value could exceed the maximum short-circuit of a panelboard rating and the circuit breakers contained within the panelboard, a high-quality fuse with the ability to limit the maximum available short-circuit current should be installed in the circuit path ahead of the panelboard. The added current-limiting protection offered by the fuse allows the panelboard to be installed without fear of damage to the electrical equipment.

As electrical demands increase, the electric utility will increase the capacity of the distribution system by providing newer equipment, such as distribution transformers with lower impedance ratings. When distribution systems are updated with transformers with lower impedance ratings, the available short-circuit currents on the system increase because of these greater efficiencies of the new equipment. When utilities upgrade their distribution system, they recommend that customers who will be supplied by the newer equipment have an electrical designer or contractor reevaluate their electrical systems to determine whether they are capable of safely withstanding the higher level of short-circuit current that will be available. If the existing equipment is not capable of safely withstanding the higher short-circuit current levels, appropriate alterations as necessary must be made to ensure that the electrical system meets all requirements.

The *NEC* requires that all components in an electrical system must be protected from damaging currents due to short circuits [110.10 and 110.09].

Wrap Up

■ Master Concepts

- After the combined load for a facility has been calculated, designers complete a set of load calculations to determine more accurately the actual electrical demands that may be imposed on a facility's electrical system.
- Short-circuit amperages can be very hazardous to electrical equipment and possibly to personnel working on or around electrical equipment.
- Designers must complete calculations to evaluate and design components into the electrical system that can reduce or eliminate the hazards associated with short-circuit amperages.

■ Charged Terms

<u>C value (for conductors)</u> Multipliers that have been developed for conductors that are derived by including both the resistance and the impedance of a conductor (X/R) installed in electrical systems. These multipliers are used in short-circuit calculations and result in calculation of more accurate short-circuit current values.

<u>counter electromotive force (CEMF)</u> An induced voltage that results in a force opposite in direction to the applied voltage; in AC circuits with magnetic properties (such as motors and transformers), this induced voltage can cause the circuit current to lag the applied voltage, resulting in lower power factor values.

<u>demand factor</u> The ratio of power consumed by a system at any one time to the maximum power that would be consumed if the entire load connected to the system were to be operating at the same time.

<u>load calculations</u> A set of calculated values that determine the demand factor for a system and that reflects a more true value of power utilized at any one time compared to calculated values determined during design.

<u>point-to-point method</u> A calculation method to determine the available short-circuit current values at any point in a system.

<u>short-circuit calculations</u> A set of calculated values that determine available short-circuit currents.

<u>transformer impedance rating</u> A voltage drop rating for a transformer given in a percentage (Z) of the full load voltage.

■ Check Your Knowledge

1. Load calculations are completed to determine load demand for:
   A. the entire electrical system.
   B. specialized electrical equipment only.
   C. lighting loads only.
   D. general electrical equipment only.

2. Demand factors are applied to certain electrical loads to determine more accurately:
   A. which loads should be included in total load calculations.
   B. electrical system demands based on future requirements.
   C. total electrical demand of the system.
   D. maximum capacity of designed loads.

3. When designing feeder and branch-circuit conductors that will serve continuous loads, the conductors must be based on what percentage of the calculated load?
   A. 100 percent
   B. 125 percent
   C. 150 percent
   D. 200 percent

4. If a 3-phase, 4-wire, 120/208-volt panelboard will serve a continuous load of 44.9 kVA and a noncontinuous load of 30.6 kVA, what is the total calculated load for the panelboard?
   A. 64.3 kVA
   B. 75.5 kVA
   C. 86.7 kVA
   D. 122.3 kVA

5. What is the total calculated demand for a 3-phase, 4-wire, 120/208-volt panelboard serving both a lighting load of 23.4 kVA and a general electrical load of 16.2 kVA?
   A. 39.6 kVA
   B. 42.3 kVA
   C. 49.5 kVA
   D. 63.0 kVA

6. Which of the following is NOT a typical short-circuit rating for electrical equipment?
   A. 10,000 short-circuit amperes
   B. 15,000 short-circuit amperes
   C. 42,000 short-circuit amperes
   D. 65,000 short-circuit amperes

7. What is the minimum copper grounding electrode conductor size for an electrical service fed with 400-kcmil service entrance conductors?
   A. Size 6 AWG
   B. Size 4 AWG
   C. Size 2 AWG
   D. Size 1/0 AWG

8. What is the minimum copper grounding electrode conductor size for a service fed with two 1100-kcmil conductors per phase?
   A. Size 1/0 AWG
   B. Size 2/0 AWG
   C. Size 3/0 AWG
   D. Size 4/0 AWG

9. What is the total load demand in amperes for a group of motors with the following ratings: M1 = 15.2 A, M2 = 15.2 A, M3 = 54.0 A, M4 = 68.0 A?
   A. 152.4 A
   B. 169.4 A
   C. 381 A
   D. 228.75 A

10. What is the short-circuit motor contribution for a 125–hp, 3-phase, 460-volt motor with a starting amperage six times normal full-load running amperage?
    A. 195 A
    B. 234 A
    C. 390 A
    D. 963 A

You Are the Designer

Apply the knowledge you have gained from this previous chapter to your own electrical design. In this section you will:
- Complete a set of load calculations
- Complete all required short-circuit calculations
- Determine the size of the main switchboard that will serve your design

■ About Your Project

To complete your task, you must know the following details about your project:
- All panel schedules (previously designed)
- Single-line diagram (previously designed)

■ Resources

To develop this part of your design, you need the following resources:
- Completed E_3 (included in the text and on the *Student Resource CD-ROM*)

■ Get to Work

At this stage in the design process, you compile several previously designed components to finalize the plan sheet that illustrates the single-line diagram, panel schedules, and load and short-circuit calculations. For each panel schedule, classify the loads by type (e.g., general electrical, specialized electrical, lighting) and complete a set of load calculations as outlined in the chapter. Apply all demand factors as necessary:
- For general electrical loads, the first 10,000 VA are calculated at 100 percent, and the remainder of the load is calculated at 50 percent.
- For specialized electrical loads, use 100 percent of the load—except for loads with motors, which require 125 percent of the highest-rated motor in the group plus 100 percent for all remaining motors.
- For lighting loads, apply an additional 25 percent of the total load for the long continuous load factor.

Compute the final totals and add them to the Main Distribution section of your plan (see E_3) to calculate the total amperage. Reference Table 5-1 to determine the minimum size switchboard necessary for your project. Illustrate the size of the main disconnect on your single-line diagram and list the number of poles and main fuse size.

Lighting loads will be on for three hours or more, therefore they are considered continuous loads and an additional 25 percent must be added to the designed load, as per *NEC* 215.2(A)(1).

Next, complete your single-line diagram by listing the method and size of the grounding electrode conductor for the main switchboard (based on NEC Table 250.66). Remember that if the service entrance conductors are in parallel, the grounding electrode conductor is based on the size of the total circular mil area for both conductors. In the sample completed plan, the grounding electrode conductor size is referenced as being a size 3/0 conductor, which was obtained from NEC Table 250.66 based on the area of 1200 circular mil, which is the total of the two 600-kcmil conductors installed in parallel.

General purpose receptacle outlets are not considered to be continuous loads. For these types of loads, demand factors (reductions in designed load) can be taken.

For the sample project, the following load calculations were applied:

Panelboard P_1		Panelboard P_2		Panelboard P_3	
Lighting load	28,306 VA	Receptacle load	9000 VA	Receptacle load	(43,200 VA total)
Long continuous load (25 percent)*	7077 VA	Motor load (including 125 percent of the highest-rated motor)	53,995 VA	100 percent of first 10,000 VA	10,000 VA
Motor load (including 125 percent of the highest-rated motor)	151,500 VA			50 percent of remaining load	16,600 VA
Transformer T_1	60,818 VA			Equipment load	13,080 VA
Total volt-amperes	247,751 VA	Total volt-amperes	62,995 VA	Total volt-amperes	39,680 VA
Total amperes	298 A (480-V)	Total amperes	76 A (480-V)	Total amperes	47.8 A (480-V)

*As per NEC 210.19(A) and 215.2(A)(1), lighting loads must be calculated at 125 percent due to the continuous load application, so an additional 25 percent was added to the total.

Main Distribution	
P_1	247,751 VA
P_2	62,995 VA
P_3	39,680 VA
Motor load (including 125 percent of the highest-rated motor)	160,475 VA
Total	510,901 VA (616 A)

Next, determine the demand load for the individual loads served from the main switchboard. For example, for panelboard P_1:

1. Determine the total lighting load including the required long continuous load value:

Lighting load from panel schedule	28,306 VA
Long continuous load (25 percent of total lighting load)	+ 7,077 VA
Total lighting load	35,383 VA

2. Determine the motor load by adding 125 percent of the highest-rated motor in the group and 100 percent of all remaining motors:

125 percent of highest-rated motor (CNC Mill 2)	28,233 VA × 1.25 =	35,291 VA
100 percent of all remaining motors		+116,259 VA
Total motor load		151,550 VA

3. Determine the total volt-ampere load from the transformer:

   T_1 (from panel schedule P_1) = 60,818 VA

4. Add these values together to determine total load:

   35,383 VA + 151,550 VA + 60,818 VA = 247,751 VA

5. Use Ohm's law to determine the total calculated load in amperes:

$$\text{Amperes } (I) = \frac{VA}{\text{voltage } (E) \times 1.73}$$

$$= \frac{247,751 \text{ VA}}{480 \text{ V} \times 1.73}$$

$$= 298 \text{ A}$$

Answer: The total demand load for P_1 is 298 A.

The 25 percent added to the highest motor in the group provides additional capacity for the motor's inrush current. When the additional percentage is added to the largest motor, this is sufficient for other motors in the group of smaller horsepower.

The same guidelines were followed for panelboard P_2. The load on panelboard P_3 includes both general purpose receptacle loads and specialized equipment loads (e.g., copy machines, vending machines). The total values for each panelboard were added together in addition to the motor loads served directly from the main switchboard to determine the total load for the main switchboard.

The next step is to determine the total connected load for a main switchboard. For this project:

1. Add the total loads for each panelboard and the motor load served from the main switchboard (including 125 percent of the highest-rated motor):

Panelboard P_1	247,751 VA
Panelboard P_2	62,995 VA
Panelboard P_3	39,680 VA
Motor load	+ 160,475 VA
Total panelboard load	510,901 VA

   Calculate the load in amperes using Ohm's law:

   $$\text{Amperes } (I) = \frac{VA}{\text{Voltage } (E) \times 1.73}$$

   $$= \frac{510{,}901 \text{ VA}}{480 \text{ V} \times 1.73}$$

   $$= 579 \text{ A}$$

Answer: The total connected load for this switchboard is 616 A.

Once you calculate this total load, the next step is to size the main switchboard using the standard National Electrical Manufacturers Association (NEMA) panelboard sizes presented in Table 5-1. The minimum size switchboard necessary to serve this facility is 800 A.

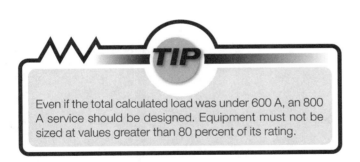

TIP: Even if the total calculated load was under 600 A, an 800 A service should be designed. Equipment must not be sized at values greater than 80 percent of its rating.

Now that the load calculations are complete, you must perform the required short-circuit calculations. Carefully complete each calculation as outlined in the text and illustrated on completed plan sheet E_3. Enter each result in the appropriate area on your single-line diagram. Additionally, you should review the standard short-circuit rating for devices and illustrate the main fuse disconnect and the standard value for the main switchboard. For example, for the completed project it was determined that the available short-circuit amperage exceeded 22,000 short-circuit A; therefore, it was necessary for the equipment to be rated at the next higher available rating of 42,000 short-circuit A.

To determine the short-circuit amperage for distribution panel P_1, you must start your calculations in step 1 with the maximum available short-circuit amperage at the main switchboard. The available short-circuit amperage at this panelboard (and other panelboards served by the main switchboard) will be derated because of the additional impedance added by conductor sizes, lengths, and raceway used. Follow the same procedure for the other panelboards. Note that on panelboard P_3, there is no motor contribution, so no additional amperage was added.

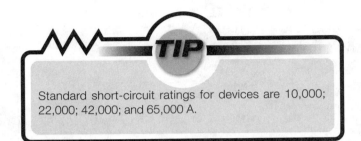

Standard short-circuit ratings for devices are 10,000; 22,000; 42,000; and 65,000 A.

CHAPTER 7

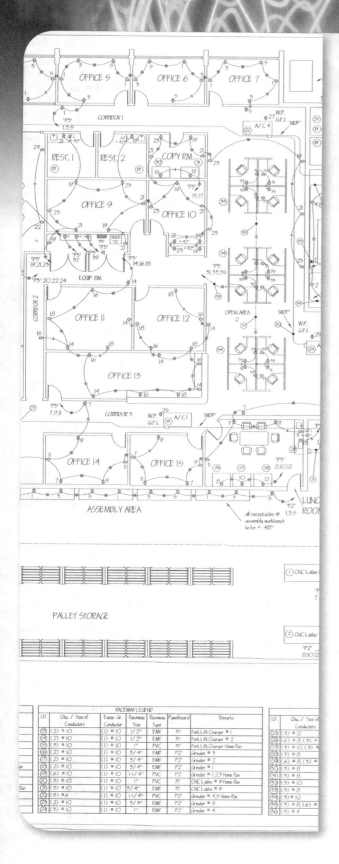

Electrical Plan Review

Chapter Outline

- **Introduction**
- **Electrical Plan Review Process**
- **Electrical Plan Review Checklist**

Learning Objectives

- Define the stages of the electrical plan review process required to obtain a building permit.

- Recognize the components in a standard electrical plan review checklist and the various components of general and specialized electrical requirements, lighting systems, and distributions systems that must be reviewed and approved before submission to a local building division.

- Comprehend how to organize a design plan and relevant documentation into a complete set of construction plans.

Introduction

In the final stage of electrical plan design, the designer, facility owner or occupant, and local building authorities thoroughly review the plan for any errors or omissions. After making any necessary corrections and approving the plan, building permits can be issued so that the bid proposal and construction phase of the project can begin (see **FIGURE 7-1**).

Electrical Plan Review Process

The review process goes through several stages:
1. The electrical designer completes an in-house review, carefully reviewing each aspect of the plan and making any necessary changes.
2. The electrical designer compiles and organizes the plan into a final complete set of **construction plans** that can then be made available to all relevant parties. The number of plan sheets varies based on the scope of the project. At a minimum, each set will include the following components:
   - E_1: Electrical floor plan
   - E_2: Lighting plan
   - E_3: Panel schedule(s), **single-line diagram**, and load and short-circuit calculations
3. This set is presented to the appropriate parties (i.e., building owners or occupants), who then can convey their final comments, additions, or changes.
4. A **plan check engineer** at the local building division reviews the plan for any errors or omissions and verifies that all the required documentation (e.g., power plan, lighting plan, and required documentation for lighting energy requirements) has been submitted. Plan check engineers use a standard checklist to review the design and submit a corrections list to the designer, noting any changes that must be made to obtain a building permit.
5. If necessary, the designer makes any required changes to the plan and resubmits it to the local building division for further review.
6. Once the plan has been approved, the local building authority issues a building permit.

Organize the documentation for a project in a folder labeled with the project title and include any questions, revisions, and reference information. Accurate, current, and well-organized documentation is of utmost importance when reference needs to be made for any part of the design.

Before submission, confirm the number of sets required. Many jurisdictions require a minimum of four complete sets.

Electrical Plan Review Checklist

The following sections should all be reviewed thoroughly by the electrical designer to verify that the

Many jurisdictions review a plan and submit a corrections list in approximately 30 days. Incomplete submissions will likely cause delays, and the plan may need to be resubmitted.

FIGURE 7-1 Once the electrical review process is complete, the construction phase can begin.

plan is complete and that all the components of the design are correct. Please note that this checklist does not ensure that the plan adheres to the proper codes and regulations; this check should have already been completed during the design process.

General and Specialized Electrical Requirements

The designer must review the electrical design, checking for inclusion and accuracy of the following components:

- Receptacles
- Circuitry
  - Branch-circuit numbers (referenced at each receptacle)
  - Home run raceway designation
- Conductors or conduits (the proper quantity for each raceway)
- Specialty circuits for equipment (e.g., shop machinery)
- Raceway legend
- Panel schedule(s) (*Note:* Provisions listed are for four individual panel schedules; sections for panelboards can be added or deleted as required):
  - Consistent and correct title (e.g., P_1, L_1)
  - Proper voltage and phase
  - Size of the main overcurrent protection
  - Circuit breaker sizes
  - Branch-circuit overcurrent device size and number of poles
  - Totals in amperage and volt-amperes or kVA
  - Ten percent load balancing across phases, achieved as best as possible
  - Required calculations for any continuous loads in the panel schedule (included in the amperage and volt-ampere totals)

TIP

A provision of 10 percent load balancing is provided to verify that calculations have been performed so that panel schedule phase totals are balanced to within 10 percent across the phases.

Lighting System

The designer must review the lighting plan, checking for inclusion and accuracy of the following components:

- Lighting fixtures (properly identified at each installed location and cross-referenced to the lighting fixture schedule)
- Control devices
- Energy management controls, if applicable (*Note:* For any lighting energy management controls, additional documentation describing installation requirements and connection diagrams is required.)
- Circuitry
  - Lighting branch-circuit numbers
  - Home run raceway designation
- Lighting fixture schedule
  - Fixture list with appropriate alphabetical designation
  - Fixture type
  - Lamp type
  - Mounting method
  - Description
  - Manufacturer catalog number
- Energy documentation with the appropriate lighting energy calculations

Distribution System

The designer must review the single-line diagram, checking for overall completeness of the main switchboard and confirming that all the following information is provided in the correct location, appropriately illustrated, and properly annotated:

- Switchboard size (in amperage, voltage, and number of phases)
- Disconnecting means type and size (fuse or circuit breaker, size of the protection in amperage, and required number of poles)
- Service entrance conductor and raceway type and size (*Note:* With underground raceways, these values may be determined by the serving utility.)
- System grounding electrode conductor and raceway type and size and system grounding method
- Load calculations
- Short-circuit calculations

TIP

The majority of errors in an electrical design are found in the single-line diagram, so review this document very carefully.

TIP

Ensure that the same naming convention that is used for panelboard identification in a panel schedule (e.g., P_1, L_1) is used on the single-line diagram. Inconsistent labeling could lead to serious safety issues during service or repairs.

The designer must review the single-line diagram, checking for inclusion and accuracy of the following components of the distribution system:

- Feeder conductor type and size
- Branch-circuit conductor type and size
- Raceway type and size
- Grounding conductor type and size
- Distribution transformers
  - Size (in kVA)
  - Primary and secondary voltages
  - Number of phases
  - Primary and secondary conductor sizes
  - Primary and secondary overcurrent protection sizes, if required
  - Raceway types and sizes
  - Grounding method(s)
  - Grounding conductor types and sizes
- Load amperage

Wrap Up

■ Master Concepts

- The electrical designer must perform an in-house review of a completed plan, checking for any errors or omissions.
- A comprehensive predeveloped checklist should be used for the in-house review to ensure that all required design details are present on the plan.
- Once a plan has been reviewed by the electrical designer, it is made available to the owner or occupant of the facility for any revisions.
- The final approved plan is submitted to the local building department for review and approval and is released to any parties who may be asked to submit a bid proposal for the project.

■ Charged Terms

<u>construction plans</u> A complete set of building plans which can be made available to all relevant parties providing an estimate for or performing work on a given project.

<u>plan check engineer</u> An engineer who performs building and plan examinations for construction or alteration of industrial, commercial, and residential structures and determines compliance with applicable codes, laws, and regulations.

<u>single-line diagram</u> A simple diagram (also called one-line diagram) that illustrates all the information and requirements of an electrical distribution system.

■ Check Your Knowledge

1. The purpose of the in-house review is to:
    - A. check the design for any errors or omissions.
    - B. allow the plan check engineer to compare the submitted design to a template.
    - C. verify that the design meets all *NEC* requirements.
    - D. track revisions made to a project.

2. The most common place for errors to occur in an electrical design is the:
    - A. power plan.
    - B. lighting plan.
    - C. single-line diagram.
    - D. raceway legend.

3. An in-house review process should include the:
    - A. power plan.
    - B. lighting plan.
    - C. single-line diagram.
    - D. All of the above

4. Once a plan has been reviewed by the electrical designer, it is made available to the _____ so that any additions or deletions can either be redesigned or omitted.

   A. Local building department
   B. Owner or occupant of the facility
   C. Local serving utility company
   D. Contractors who may be providing a bid proposal for the project

5. Once a plan is submitted to a local building division, a building permit or corrections list is usually returned within:

   A. 30 days.
   B. 60 days.
   C. 3 months.
   D. 6 months.

You Are the Designer

Apply the knowledge you have gained from this previous chapter to your own electrical design. In this section you will:
- Thoroughly review your completed design
- Implement any necessary changes to your design
- Prepare and compile the design and required documentation for submission to a local building division

■ About Your Project

To complete your task, you must gather the following details from your project:
- Your completed plan sheets: E_1, E_2, and E_3
- All panel schedules
- Lighting fixture schedule

■ Resources

To develop this final part of your design you need the following resources:
- Electrical Plan Review Checklist from the *Student Resource CD-ROM*

■ Get to Work

The last step of your project is to go through a checklist like the one used by local building officials during the plan approval process to verify that all the necessary documentation is complete. Note that this process does not confirm that the design adheres to applicable codes (you should have done that during previous stages of the design). For each item listed in the Plan Check document, check "OK" or "REV" (which stands for "revision required") for any incorrect or omitted items. If necessary, make the appropriate revisions and check your plan again. When you have marked all items "OK," you have successfully completed your design project. The following pages show a sample checklist.

SAMPLE ELECTRICAL PLAN REVIEW CHECKLIST

SINGLE-LINE DIAGRAM	OK	REV	NOTES
MAIN SWITCH GEAR			
Switchboard Main Disconnect Size			
Service Entrance Conductor Size			
Service Entrance Conduit Size(s)			
System Grounding Electrode Conductor Size			
Short-Circuit Calculations			
Load Calculations			
DISTRIBUTION / SUB FEEDS			
Feeder Conductor Sizes			
Raceway Sizes			
Raceway Distances			
Equipment Grounding Conductor			
Load Amps			
TRANSFORMERS			
kVa Size / Primary and Secondary Voltages			
Primary and Secondary Conductors Sizes			
Overcurrent Protection Primary and Secondary			
Raceway Sizes			
Raceway Distances			
System Grounding Conductor Sizes			
Short-Circuit Values			
PANEL SCHEDULES			
PANEL 1			
Distribution Panel Identification			
Voltage/Phase			
Main Overcurrent Protection			
Branch Circuit Breaker Sizes and Number of Poles			
Individual Phase Totals			
Long Continuous Load Value			
Balanced Loads Across Each Phase (10%)			
Totals / kVa / Amps			
Short-Circuit Values Listed on Single-Line Diagram			
PANEL 2			
Distribution Panel Identification			
Voltage/Phase			
Main Overcurrent Protection			
Branch-Circuit Breaker Sizes and Number of Poles			
Individual Phase Totals			
Long Continuous Load Value			
Balanced Loads Across Each Phase (10%)			
Totals / kVa / Amps			
Short-Circuit Values Listed on Single-Line Diagram			

SINGLE-LINE DIAGRAM	OK	REV	NOTES
PANEL 3			
Distribution Panel Identification			
Voltage / Phase			
Main Overcurrent Protection			
Branch-Circuit Breaker Sizes and Number of Poles			
Individual Phase Totals			
Long Continuous Load Value			
Balanced Loads Across Each Phase (10%)			
Totals / kVa / Amps			
Short-Circuit Values Listed on Single-Line Diagram			
PANEL 4			
Distribution Panel Identification			
Voltage/Phase			
Main Overcurrent Protection			
Branch-Circuit Breaker Sizes and Number of Poles			
Individual Phase Totals			
Long Continuous Load Value			
Balanced Loads Across Each Phase (10%)			
Totals / kVa / Amps			
Short-Circuit Values Listed on Single-Line Diagram			
LIGHTING FIXTURE SCHEDULE			
Fixture Identification			
Fixture Type			
Lamp Information			
Lamp Wattage and Type			
Mounting Method			
Description and Variations			
Manufacturer Catalog Number			
POWER PLAN			
Receptacles			
Circuitry / Home Runs			
Conductor / Raceways			
Shop Machinery			
a) Raceway Sizes			
b) Conductor Sizes			
c) Circuitry / Home Runs			
d) Raceway Legend			
LIGHTING PLAN			
Fixture Identification			
Switching Methods			
Energy Controls			
Circuitry / Home Runs			
Project Notes			

Glossary

120/208-volt, 3-phase, 4-wire, wye system A distribution system generated with three individual sine waves separated by 120 electrical degrees that are identified as phases A, B, and C. One leg of each of the three phase coils is electrically connected to the others at a common point, forming a wye, which when grounded, becomes the fourth wire (or neutral) in the system. This allows for each of the three individual phase voltages to supply 120 volts to the grounded point, while the line voltage across each of the phases produces 208 volts. The line-to-line voltages can supply both 208-volt 3-phase and 208-volt single-phase. Because the three individual phases each can supply 120 volts, this system is commonly used in commercial office applications where 120 volts is desired because the 120-volt loads can be balanced across each of the three phases.

277/480-volt, 3-phase, 4-wire, wye system A distribution system generated with three individual sine waves separated by 120 electrical degrees that are identified as phases A, B, and C. One leg of each of the three phase coils is electrically connected to the others at a common point, forming a wye, which when grounded becomes the fourth wire (or neutral) in the system. This allows for each of the three individual phase voltages to supply 277 volts to the grounded point, while the line voltage across each of the phases produces 480 volts. The line-to-line voltages can supply both 480-volt 3-phase and 480-volt single-phase. This system is commonly used in commercial applications where 480 volts is required for machinery loads and in applications to serve 277-volt lighting loads.

A/B switching method A dual switching method to control lighting that reduces the connected lighting load by at least 50 percent, maintains reasonably uniform illumination, and helps meet mandated lighting energy requirements.

American National Standards Institute (ANSI) An organization that oversees the development of standards created by manufacturers throughout the industry to promote safety and other standards.

area category method A lighting design method that designers can use to meet mandated lighting energy codes. In this method, the designer assigns a maximum allowable watts per square foot level to specifically define areas to provide adequate luminance for the primary function of the occupancy type. This method allows for maximum power levels in any one space to be exceeded when other spaces can be designed at lower allowed levels. The net result is that total power used does not exceed the calculated maximum allowed value. (*See also* complete building method.)

balanced distribution An electrical distribution system in which the ungrounded conductors carry equal currents. In distribution systems that also include a grounded conductor, the grounded conductor will carry the imbalance of the currents in the ungrounded conductors.

branch circuit The circuit conductors between the final overcurrent device protecting the circuit and the outlet(s) [100].

bypass switch A switch installed to override any automatic lighting shut-off device (e.g., time clock).

C value (for conductors) Multipliers that have been developed for conductors that are derived by including both the resistance and the impedance of a conductor (X/R) installed in electrical systems. These multipliers are used in short-circuit calculations and result in calculation of more accurate short-circuit current values.

color rendering index (CRI) The ability of a lighting source to correctly represent an illuminated object in relation to natural daylight.

complete building method A lighting design method used to meet mandated lighting energy codes by calculating the maximum allowable lighting power for a facility based on a maximum allowable watts per square foot value; a basic design method not suitable for areas with specialty lighting. (*See also* area category method.)

computer-aided design (CAD) The use of computers and design software to aid in the design of drawings, objects, shapes, and other items.

construction plans A complete set of building plans which can be made available to all relevant parties providing an estimate for or performing work on a given project.

continuous load A load where the maximum current is expected to continue for three hours or more [100].

counter electromotive force (CEMF) An induced voltage that results in a force opposite in direction to the applied voltage; in AC circuits with magnetic properties (such as motors and transformers), this induced voltage can cause the circuit current to lag the applied voltage, resulting in lower power factor values.

demand factor The ratio of power consumed by a system at any one time to the maximum power that would be consumed if the entire load connected to the system were to be operating at the same time.

effective grounding path A grounding path of low resistance that ensures, either through raceway methods or additional wiring methods, that the operation of protective devices will occur to isolate a faulted system and thus protect personnel from the dangers of electrical shock or explosion.

Electrical Apparatus and Service Association (EASA) An organization that provides information and education about sales, service, and maintenance materials for motors, generators, and other electromechanical equipment.

equipment grounding conductor The conductive path installed to connect normally non-current-carrying metal parts of equipment together and to the system grounded conductor, to the grounding electrode conductor, or to both [100].

equipment list A developed table that lists details about the specialized equipment that is to be incorporated into a design plan.

foot candle (fc) A measurement of illumination intensity. One foot candle is the intensity of light on a surface 1 foot from a lighting source of 1 candlepower.

general purpose branch circuit A branch circuit that supplies two or more receptacles or outlets for lighting and appliances [100].

ground fault A condition in which high levels of current could flow when an ungrounded conductor accidentally comes in contact with a grounded reference. (*See also* short circuit.)

grounding electrode A conducting object through which a direct connection to earth is established [100].

grounding electrode conductor A conductor used to connect the system grounded conductor or the equipment to a grounding electrode or a point on the grounding electrode system [100].

home run The raceway designated on a plan as the one that carries branch-circuit conductors back to the serving source (such as a panelboard).

Illuminating and Engineering Society of North America (IESNA) An organization that works with manufacturers, designers, architects, consultants, electrical and building contractors, and suppliers with regard to lighting systems.

inrush current A momentary high level of amperage flowing in a circuit such as those associated with motorized equipment loads.

isolated ground (IG) An additional equipment grounding conductor that, when installed, provides for the grounding of equipment separate from a grounding method that uses an approved raceway method; typically used for electrically sensitive equipment in computer applications and medical facilities. When isolated grounding is provided through the use of receptacles, the receptacle must be identified on the design plan as "IG."

lighting branch circuit A branch circuit that serves only lighting.

lighting energy-saving devices Devices such as multiple switches, time clocks, and occupancy sensors that can achieve lighting energy savings.

lighting fixture schedule A document included with the lighting design plan that lists the specific information for the lighting fixtures in a facility.

lighting system Components such as branch circuits, switching devices, and energy-saving devices that are associated with lighting fixtures and their control.

load calculations A set of calculated values that determine the demand factor for a system and that reflects a more true value of power utilized at any one time compared to calculated values determined during design.

lumens per watt (lm/W) The ratio of light output (lumens) to input power (watts).

luminance level The amount of light projected on a work surface.

main disconnecting means The main device that disconnects the supply conductors from all sources of supply.

manufacturer electrical specification sheet Information provided by a manufacturer that lists specific details about the product; these specification sheets are often used to obtain information about motorized equipment and lighting fixtures.

multiwire branch circuit A branch circuit that consists of two or more ungrounded conductors that have a voltage between them, and a grounded conductor that has equal voltage between it and each ungrounded conductor of the circuit and that is connected to the neutral or grounded conductor of the system [100].

National Electrical Code (NEC) Regulatory code published by the National Fire Protection Association (NFPA); also known as NFPA 70.

National Electrical Manufacturers Association (NEMA) A trade association that provides standards for the electrical manufacturing industry including the generation, transmission and distribution, control, and end use of electricity.

occupancy sensor A device that detects the presence of personnel in a space by passive infrared or ultrasonic methods; when used in a lighting system, the sensor switches lighting fixtures on and off as occupants enter or exit the space to help save energy.

overload Operation of equipment in excess of normal, full-load rating, or of a conductor in excess of rated ampacity that, if it persists for a sufficient length of time, would cause damage or dangerous overheating. A fault, such as a short circuit or ground fault, is not an overload [100].

panel schedule An illustration of key panelboard information showing how branch circuitry is distributed, number of phases, voltage, and size in amperage; panel schedules are completed by hand calculation or by computer software.

plan check engineer An engineer who performs building and plan examinations for construction or alteration of industrial, commercial, and residential structures and determines compliance with applicable codes, laws, and regulations.

point-to-point method A calculation method to determine the available short-circuit current values at any point in a system.

programmable lighting controllers Microprocessor-based lighting controllers that can be programmed; their use results in greater lighting energy savings.

raceway An enclosed channel of metal or nonmetallic materials designed expressly for holding wires, cables, or busbars [100].

raceway legend A table developed by an electrical designer that illustrates information about raceways installed for a project.

receptacle outlet An outlet where one or more receptacles are installed [100].

reflected ceiling plan A plan that illustrates only the location of lighting fixtures and the ceiling type in which they are to be installed.

separately derived system A premises wiring system whose power is derived from a source of electrical energy or equipment other than a service. Such systems have no direct electrical connection, including a solidly connected grounded circuit conductor, to supply conductors originating in another system [100].

service entrance conductors The conductors from the service entry point to the service main disconnecting means. These can be installed as part of an overhead-type installation or in underground conduits.

short circuit A dangerous condition in which circuit conductors contact each other and reduce the intended ohmic resistance of the circuit; often referred to as a line-to-line or line-to-neutral short. (*See also* ground fault.)

short-circuit calculations A set of calculated values that determine available short-circuit currents.

single-line diagram A simple diagram (also called one-line diagram) that illustrates all the information and requirements of an electrical distribution system.

spacing criteria Mounting height ratios provided by the lighting fixture manufacturer used in calculations to determine proper mounting distances between lighting fixtures (also called row and column spacing criteria).

step-down transformer A transformer that delivers a different utilization voltage; typically used in commercial applications to lower a 480/277-volt service to 120/208 volts for office spaces.

switchboard An electrical cabinet, or cabinets (depending on the electrical requirements), that has provisions for the service entrance method, utility metering, and overcurrent protective devices that serve distributions to equipment.

transformer impedance rating A voltage drop rating for a transformer given in a percentage (Z) of the full load voltage.

utility metering equipment The components that make up the parts of an electrical cabinet used solely for the purposes of utility metering, such as kilowatt-hours/demand, meter testing points, and current transformers.

voltage drop A loss of voltage on a conductor resulting from the length of the conductor, its resistance, and the amperage imposed on the conductor.

Appendix

NEC Tables

Table 220.44	Demand Factors for Non-Dwelling Receptacle Loads	144
Table 250.122	Minimum Size Equipment Grounding Conductors for Grounding Raceway and Equipment	144
Table 250.66	Grounding Electrode Conductor for Alternating-Current Systems	145
Table 310.16	Allowable Ampacities of Insulated Conductors Rated 0 Through 2000 Volts, 60°C Through 90°C (140°F Through 194°F), Not More Than Three Current-Carrying Conductors in Raceway, Cable, or Earth (Directly Buried), Based on Ambient Temperature of 30°C (86°F)	146
Table 430.247	Full-Load Currents in Amperes, Direct-Current Motors	148
Table 430.248	Full-Load Currents in Amperes, Single-Phase Alternating-Current Motors	149
Table 430.250	Full-Load Current, Three-Phase Alternating-Current Motors	150
Table 430.52	Maximum Rating or Setting of Motor Branch-Circuit Short-Circuit and Ground-Fault Protective Devices	151
Table 450.3(B)	Maximum Rating or Setting of Overcurrent Protection for Transformers 600 Volts and Less (as a Percentage of Transformer-Rated Current)	151
Table 4	Dimensions and Percent Area of Conduit and Tubing (Areas of Conduit or Tubing for the Combinations of Wires Permitted in Table 1, Chapter 9) (partial)	152
Table 5	Dimensions of Insulated Conductors and Fixture Wires (partial)	153
Table 8	Conductor Properties	154
Table C1	Maximum Number of Conductors or Fixture Wires in Electrical Metallic Tubing (EMT)	156

NEC TABLE 220.44 Demand Factors for Non-Dwelling Receptacle Loads

Portion of Receptacle Load to Which Demand Factor Applies (Volt-Amperes)	Demand Factor (%)
First 10 kVA or less at	100
Remainder over 10 kVA at	50

Source: NEC® Handbook, NFPA, Quincy, MA, 2008, Table 220.44

NEC TABLE 250.122 Minimum Size Equipment Grounding Conductors for Grounding Raceway and Equipment

Rating or Setting of Automatic Overcurrent Device in Circuit Ahead of Equipment, Conduit, etc., Not Exceeding (Amperes)	Size (AWG or kcmil)	
	Copper	Aluminum or Copper-Clad Aluminum*
15	14	12
20	12	10
30	10	8
40	10	8
60	10	8
100	8	6
200	6	4
300	4	2
400	3	1
500	2	1/0
600	1	2/0
800	1/0	3/0
1000	2/0	4/0
1200	3/0	250
1600	4/0	350
2000	250	400
2500	350	600
3000	400	600
4000	500	800
5000	700	1200
6000	800	1200

Note: Where necessary to comply with 250.4(A)(5) or (B)(4), the equipment grounding conductor shall be sized larger than given in this table.
*See installation restrictions in 250.120.

Source: NEC® Handbook, NFPA, Quincy, MA, 2008, Table 250.122

NEC TABLE 250.66 Grounding Electrode Conductor for Alternating-Current Systems

Size of Largest Ungrounded Service-Entrance Conductor or Equivalent Area for Parallel Conductors[a] (AWG/kcmil)		Size of Grounding Electrode Conductor (AWG/kcmil)	
Copper	Aluminum or Copper-Clad Aluminum	Copper	Aluminum or Copper-Clad Aluminum[b]
2 or smaller	1/0 or smaller	8	6
1 or 1/0	2/0 or 3/0	6	4
2/0 or 3/0	4/0 or 250	4	2
Over 3/0 through 350	Over 250 through 500	2	1/0
Over 350 through 600	Over 500 through 900	1/0	3/0
Over 600 through 1100	Over 900 through 1750	2/0	4/0
Over 1100	Over 1750	3/0	250

Notes:
1. Where multiple sets of service-entrance conductors are used as permitted in 230.40, Exception No. 2, the equivalent size of the largest service-entrance conductor shall be determined by the largest sum of the areas of the corresponding conductors of each set.
2. Where there are no service-entrance conductors, the grounding electrode conductor size shall be determined by the equivalent size of the largest service-entrance conductor required for the load to be served.

[a] This table also applies to the derived conductors of separately derived ac systems.
[b] See installation restrictions in 250.64(A).

Source: NEC® Handbook, NFPA, Quincy, MA, 2008, Table 250.66

NEC TABLE 310.16 Allowable Ampacities of Insulated Conductors Rated 0 Through 2000 Volts, 60°C Through 90°C (140°F Through 194°F), Not More Than Three Current-Carrying Conductors in Raceway, Cable, or Earth (Directly Buried), Based on Ambient Temperature of 30°C (86°F)

	Temperature Rating of Conductor [See Table 310.13(A).]						
	60°C (140°F)	75°C (167°F)	90°C (194°F)	60°C (140°F)	75°C (167°F)	90°C (194°F)	
	Types TW, UF	Types RHW, THHW, THW, THWN, XHHW, USE, ZW	Types TBS, SA, SIS, FEP, FEPB, MI, RHH, RHW-2, THHN, THHW, THW-2, THWN-2, USE-2, XHH, XHHW, XHHW-2, ZW-2	Types TW, UF	Types RHW, THHW, THW, THWN, XHHW, USE	Types TBS, SA, SIS, THHN, THHW, THW-2, THWN-2, RHH, RHW-2, USE-2, XHH, XHHW, XHHW-2, ZW-2	
Size AWG or kcmil	COPPER			ALUMINUM OR COPPER-CLAD ALUMINUM			Size AWG or kcmil
18	—	—	14	—	—	—	—
16	—	—	18	—	—	—	—
14*	20	20	25	—	—	—	—
12*	25	25	30	20	20	25	12*
10*	30	35	40	25	30	35	10*
8	40	50	55	30	40	45	8
6	55	65	75	40	50	60	6
4	70	85	95	55	65	75	4
3	85	100	110	65	75	85	3
2	95	115	130	75	90	100	2
1	110	130	150	85	100	115	1
1/0	125	150	170	100	120	135	1/0
2/0	145	175	195	115	135	150	2/0
3/0	165	200	225	130	155	175	3/0
4/0	195	230	260	150	180	205	4/0
250	215	255	290	170	205	230	250
300	240	285	320	190	230	255	300
350	260	310	350	210	250	280	350
400	280	335	380	225	270	305	400
500	320	380	430	260	310	350	500
600	355	420	475	285	340	385	600
700	385	460	520	310	375	420	700

	Temperature Rating of Conductor [See Table 310.13(A).]						
	60°C (140°F)	75°C (167°F)	90°C (194°F)	60°C (140°F)	75°C (167°F)	90°C (194°F)	
	Types TW, UF	Types RHW, THHW, THW, THWN, XHHW, USE, ZW	Types TBS, SA, SIS, FEP, FEPB, MI, RHH, RHW-2, THHN, THHW, THW-2, THWN-2, USE-2, XHH, XHHW, XHHW-2, ZW-2	Types TW, UF	Types RHW, THHW, THW, THWN, XHHW, USE	Types TBS, SA, SIS, THHN, THHW, THW-2, THWN-2, RHH, RHW-2, USE-2, XHH, XHHW, XHHW-2, ZW-2	
Size AWG or kcmil	COPPER			ALUMINUM OR COPPER-CLAD ALUMINUM			Size AWG or kcmil
750	400	475	535	320	385	435	750
800	410	490	555	330	395	450	800
900	435	520	585	355	425	480	900
1000	455	545	615	375	445	500	1000
1250	495	590	665	405	485	545	1250
1500	520	625	705	435	520	585	1500
1750	545	650	735	455	545	615	1750
2000	560	665	750	470	560	630	2000
CORRECTION FACTORS							
Ambient Temp. (°C)	For ambient temperatures other than 30°C (86°F), multiply the allowable ampacities shown above by the appropriate factor shown below.						Ambient Temp. (°F)
21–25	1.08	1.05	1.04	1.08	1.05	1.04	70–77
26–30	1.00	1.00	1.00	1.00	1.00	1.00	78–86
31–35	0.91	0.94	0.96	0.91	0.94	0.96	87–95
36–40	0.82	0.88	0.91	0.82	0.88	0.91	96–104
41–45	0.71	0.82	0.87	0.71	0.82	0.87	105–113
46–50	0.58	0.75	0.82	0.58	0.75	0.82	114–122
51–55	0.41	0.67	0.76	0.41	0.67	0.76	123–131
56–60	—	0.58	0.71	—	0.58	0.71	132–140
61–70	—	0.33	0.58	—	0.33	0.58	141–158
71–80	—	—	0.41	—	—	0.41	159–176

*See 240.4(D).

Source: NEC® Handbook, NFPA, Quincy, MA, 2008, Table 310.16

NEC TABLE 430.247 Full-Load Current in Amperes, Direct-Current Motors
The following values of full-load currents* are for motors running at base speed.

Horsepower	Armature Voltage Rating*					
	90 Volts	120 Volts	180 Volts	240 Volts	500 Volts	550 Volts
¼	4.0	3.1	2.0	1.6	—	—
⅓	5.2	4.1	2.6	2.0	—	—
½	6.8	5.4	3.4	2.7	—	—
¾	9.6	7.6	4.8	3.8	—	—
1	12.2	9.5	6.1	4.7	—	—
1½	—	13.2	8.3	6.6	—	—
2	—	17	10.8	8.5	—	—
3	—	25	16	12.2	—	—
5	—	40	27	20	—	—
7½	—	58	—	29	13.6	12.2
10	—	76	—	38	18	16
15	—	—	—	55	27	24
20	—	—	—	72	34	31
25	—	—	—	89	43	38
30	—	—	—	106	51	46
40	—	—	—	140	67	61
50	—	—	—	173	83	75
60	—	—	—	206	99	90
75	—	—	—	255	123	111
100	—	—	—	341	164	148
125	—	—	—	425	205	185
150	—	—	—	506	246	222
200	—	—	—	675	330	294

*These are average dc quantities.

Source: *NEC® Handbook*, NFPA, Quincy, MA, 2008, Table 430.247

NEC TABLE 430.248 Full-Load Currents in Amperes, Single-Phase Alternating-Current Motors

The following values of full-load currents are for motors running at usual speeds and motors with normal torque characteristics. The voltages listed are rated motor voltages. The currents listed shall be permitted for system voltage ranges of 110 to 120 and 220 to 240 volts.

Horsepower	115 Volts	200 Volts	208 Volts	230 Volts
1/6	4.4	2.5	2.4	2.2
1/4	5.8	3.3	3.2	2.9
1/3	7.2	4.1	4.0	3.6
1/2	9.8	5.6	5.4	4.9
3/4	13.8	7.9	7.6	6.9
1	16	9.2	8.8	8.0
1½	20	11.5	11.0	10
2	24	13.8	13.2	12
3	34	19.6	18.7	17
5	56	32.2	30.8	28
7½	80	46.0	44.0	40
10	100	57.5	55.0	50

Source: NEC® Handbook, NFPA, Quincy, MA, 2008, Table 430.248

NEC TABLE 430.250 Full-Load Current, Three-Phase Alternating-Current Motors

The following values of full-load currents are typical for motors running at speeds usual for belted motors and motors with normal torque characteristics.

The voltages listed are rated motor voltages. The currents listed shall be permitted for system voltage ranges of 110 to 120, 220 to 240, 440 to 480, and 550 to 600 volts.

Horsepower	Induction-Type Squirrel Cage and Wound Rotor (Amperes)							Synchronous-Type Unity Power Factor* (Amperes)			
	115 Volts	200 Volts	208 Volts	230 Volts	460 Volts	575 Volts	2300 Volts	230 Volts	460 Volts	575 Volts	2300 Volts
½	4.4	2.5	2.4	2.2	1.1	0.9	—	—	—	—	—
¾	6.4	3.7	3.5	3.2	1.6	1.3	—	—	—	—	—
1	8.4	4.8	4.6	4.2	2.1	1.7	—	—	—	—	—
1½	12.0	6.9	6.6	6.0	3.0	2.4	—	—	—	—	—
2	13.6	7.8	7.5	6.8	3.4	2.7	—	—	—	—	—
3	—	11.0	10.6	9.6	4.8	3.9	—	—	—	—	—
5	—	17.5	16.7	15.2	7.6	6.1	—	—	—	—	—
7½	—	25.3	24.2	22	11	9	—	—	—	—	—
10	—	32.2	30.8	28	14	11	—	—	—	—	—
15	—	48.3	46.2	42	21	17	—	—	—	—	—
20	—	62.1	59.4	54	27	22	—	—	—	—	—
25	—	78.2	74.8	68	34	27	—	53	26	21	—
30	—	92	88	80	40	32	—	63	32	26	—
40	—	120	114	104	52	41	—	83	41	33	—
50	—	150	143	130	65	52	—	104	52	42	—
60	—	177	169	154	77	62	16	123	61	49	12
75	—	221	211	192	96	77	20	155	78	62	15
100	—	285	273	248	124	99	26	202	101	81	20
125	—	359	343	312	156	125	31	253	126	101	25
150	—	414	396	360	180	144	37	302	151	121	30
200		552	528	480	240	192	49	400	201	161	40
250	—	—	—	—	302	242	60	—	—	—	—
300	—	—	—	—	361	289	72	—	—	—	—
350	—	—	—	—	414	336	83	—	—	—	—
400	—	—	—	—	477	382	95	—	—	—	—
450	—	—	—	—	515	412	103	—	—	—	—
500	—	—	—	—	590	472	118	—	—	—	—

*For 90 and 80 percent power factor, the figures shall be multiplied by 1.1 and 1.25, respectively.

Source: NEC® Handbook, NFPA, Quincy, MA, 2008, Table 430.250

NEC TABLE 430.52 Maximum Rating or Setting of Motor Branch-Circuit Short-Circuit and Ground-Fault Protective Devices

Type of Motor	Percentage of Full-Load Current			
	Nontime Delay Fuse[1]	Dual Element (Time-Delay) Fuse[1]	Instantaneous Trip Breaker	Inverse Time Breaker[2]
Single-phase motors	300	175	800	250
AC polyphase motors other than wound-rotor	300	175	800	250
Squirrel cage—other than Design B energy-efficient	300	175	800	250
Design B energy-efficient	300	175	1100	250
Synchronous[3]	300	175	800	250
Wound rotor	150	150	800	150
Direct current (constant voltage)	150	150	250	150

Note: For certain exceptions to the values specified, see 430.54.
[1] The values in the Nontime Delay Fuse column apply to Time-Delay Class CC fuses.
[2] The values given in the last column also cover the ratings of nonadjustable inverse time types of circuit breakers that may be modified as in 430.52(C)(1), Exception No. 1 and No. 2.
[3] Synchronous motors of the low-torque, low-speed type (usually 450 rpm or lower), such as are used to drive reciprocating compressors, pumps, and so forth, that start unloaded, do not require a fuse rating or circuit-breaker setting in excess of 200 percent of full-load current.

Source: NEC® Handbook, NFPA, Quincy, MA, 2008, Table 430.52

NEC TABLE 450.3(B) Maximum Rating or Setting of Overcurrent Protection for Transformers 600 Volts and Less (as a Percentage of Transformer-Rated Current)

Protection Method	Primary Protection			Secondary Protection (See Note 2.)	
	Currents of 9 Amperes or More	Currents Less Than 9 Amperes	Currents Less Than 2 Amperes	Currents of 9 Amperes or More	Currents Less Than 9 Amperes
Primary only protection	125% (See Note 1.)	167%	300%	Not required	Not required
Primary and secondary protection	250% (See Note 3.)	250% (See Note 3.)	250% (See Note 3.)	125% (See Note 1.)	167%

Notes:
1. Where 125 percent of this current does not correspond to a standard rating of a fuse or nonadjustable circuit breaker, a higher rating that does not exceed the next higher standard rating shall be permitted.
2. Where secondary overcurrent protection is required, the secondary overcurrent device shall be permitted to consist of not more than six circuit breakers or six sets of fuses grouped in one location. Where multiple overcurrent devices are utilized, the total of all the device ratings shall not exceed the allowed value of a single overcurrent device.
3. A transformer equipped with coordinated thermal overload protection by the manufacturer and arranged to interrupt the primary current shall be permitted to have primary overcurrent protection rated or set at a current value that is not more than six times the rated current of the transformer for transformers having not more than 6 percent impedance and not more than four times the rated current of the transformer for transformers having more than 6 percent but not more than 10 percent impedance.

Source: NEC® Handbook, NFPA, Quincy, MA, 2008, Table 450.3(B)

NEC TABLE 4 Dimensions and Percent Area of Conduit and Tubing (Areas of Conduit or Tubing for the Combinations of Wires Permitted in Table 1, Chapter 9) (partial)

Article 358 — Electrical Metallic Tubing (EMT)													
Metric Designator	Trade Size	Nominal Internal Diameter		Total Area 100%		60%		1 Wire 53%		2 Wires 31%		Over 2 Wires 40%	
		mm	in.	mm²	in.²	mm²	in.²	mm²	in²	mm²	in²	mm²	in²
16	½	15.8	0.622	196	0.304	118	0.182	104	0.161	61	0.094	78	0.122
21	¾	20.9	0.824	343	0.533	206	0.320	182	0.283	106	0.165	137	0.213
27	1	26.6	1.049	556	0.864	333	0.519	295	0.458	172	0.268	222	0.346
35	1¼	35.1	1.380	968	1.496	581	0.897	513	0.793	300	0.464	387	0.598
41	1½	40.9	1.610	1314	2.036	788	1.221	696	1.079	407	0.631	526	0.814
53	2	52.5	2.067	2165	3.356	1299	2.013	1147	1.778	671	1.040	866	1.342
63	2½	69.4	2.731	3783	5.858	2270	3.515	2005	3.105	1173	1.816	1513	2.343
78	3	85.2	3.356	5701	8.846	3421	5.307	3022	4.688	1767	2.742	2280	3.538
91	3½	97.4	3.834	7451	11.545	4471	6.927	3949	6.119	2310	3.579	2980	4.618
103	4	110.1	4.334	9521	14.753	5712	8.852	5046	7.819	2951	4.573	3808	5.901

Articles 352 and 353 — Rigid PVC Conduit (PVC), Schedule 40, and HDPE Conduit (HDPE)													
Metric Designator	Trade Size	Nominal Internal Diameter		Total Area 100%		60%		1 Wire 53%		2 Wires 31%		Over 2 Wires 40%	
		mm	in.	mm²	in.²	mm²	in.²	mm²	in²	mm²	in²	mm²	in²
12	⅜	—	—	—	—	—	—	—	—	—	—	—	—
16	½	15.3	0.602	184	0.285	110	0.171	97	0.151	57	0.088	74	0.114
21	¾	20.4	0.804	327	0.508	196	0.305	173	0.269	101	0.157	131	0.203
27	1	26.1	1.029	535	0.832	321	0.499	284	0.441	166	0.258	214	0.333
35	1¼	34.5	1.360	935	1.453	561	0.872	495	0.770	290	0.450	374	0.581
41	1½	40.4	1.590	1282	1.986	769	1.191	679	1.052	397	0.616	513	0.794
53	2	52.0	2.047	2124	3.291	1274	1.975	1126	1.744	658	1.020	849	1.316
63	2½	62.1	2.445	3029	4.695	1817	2.817	1605	2.488	939	1.455	1212	1.878
78	3	77.3	3.042	4693	7.268	2816	4.361	2487	3.852	1455	2.253	1877	2.907
91	3½	89.4	3.521	6277	9.737	3766	5.842	3327	5.161	1946	3.018	2511	3.895
103	4	101.5	3.998	8091	12.554	4855	7.532	4288	6.654	2508	3.892	3237	5.022
129	5	127.4	5.016	12748	19.761	7649	11.856	6756	10.473	3952	6.126	5099	7.904
155	6	153.2	6.031	18433	28.567	11060	17.140	9770	15.141	5714	8.856	7373	11.427

Source: NEC® Handbook, NFPA, Quincy, MA, 2008, Table 4, Chapter 9

NEC TABLE 5 Dimensions of Insulated Conductors and Fixture Wires (partial)

Type	Size (AWG or kcmil)	Approximate Diameter mm	Approximate Diameter in.	Approximate Area mm²	Approximate Area in.²
THHN, THWN, THWN-2	14	2.819	0.111	6.258	0.0097
	12	3.302	0.130	8.581	0.0133
	10	4.166	0.164	13.61	0.0211
	8	5.486	0.216	23.61	0.0366
	6	6.452	0.254	32.71	0.0507
	4	8.230	0.324	53.16	0.0824
	3	8.941	0.352	62.77	0.0973
	2	9.754	0.384	74.71	0.1158
	1	11.33	0.446	100.8	0.1562
	1/0	12.34	0.486	119.7	0.1855
	2/0	13.51	0.532	143.4	0.2223
	3/0	14.83	0.584	172.8	0.2679
	4/0	16.31	0.642	208.8	0.3237
	250	18.06	0.711	256.1	0.3970
	300	19.46	0.766	297.3	0.4608
Type: FEP, FEPB, PAF, PAFF, PF, PFA, PFAH, PFF, PGF, PGFF, PTF, PTFF, TFE, THHN, THWN, THWN-2, Z, ZF, ZFF					
THHN, THWN, THWN-2	350	20.75	0.817	338.2	0.5242
	400	21.95	0.864	378.3	0.5863
	500	24.10	0.949	456.3	0.7073
	600	26.70	1.051	559.7	0.8676
	700	28.50	1.122	637.9	0.9887
	750	29.36	1.156	677.2	1.0496
	800	30.18	1.188	715.2	1.1085
	900	31.80	1.252	794.3	1.2311
	1000	33.27	1.310	869.5	1.3478

Source: *NEC® Handbook*, NFPA, Quincy, MA, 2008, Table 5, Chapter 9

NEC TABLE 8 Conductor Properties

Size (AWG or kcmil)	Area mm²	Area Circular mils	Stranding Quantity	Stranding Diameter mm	Stranding Diameter in.	Overall Diameter mm	Overall Diameter in.	Overall Area mm²	Overall Area in.²
18	0.823	1620	1	—	—	1.02	0.040	0.823	0.001
18	0.823	1620	7	0.39	0.015	1.16	0.046	1.06	0.002
16	1.31	2580	1	—	—	1.29	0.051	1.31	0.002
16	1.31	2580	7	0.49	0.019	1.46	0.058	1.68	0.003
14	2.08	4110	1	—	—	1.63	0.064	2.08	0.003
14	2.08	4110	7	0.62	0.024	1.85	0.073	2.68	0.004
12	3.31	6530	1	—	—	2.05	0.081	3.31	0.005
12	3.31	6530	7	0.78	0.030	2.32	0.092	4.25	0.006
10	5.261	10380	1	—	—	2.588	0.102	5.26	0.008
10	5.261	10380	7	0.98	0.038	2.95	0.116	6.76	0.011
8	8.367	16510	1	—	—	3.264	0.128	8.37	0.013
8	8.367	16510	7	1.23	0.049	3.71	0.146	10.76	0.017
6	13.30	26240	7	1.56	0.061	4.67	0.184	17.09	0.027
4	21.15	41740	7	1.96	0.077	5.89	0.232	27.19	0.042
3	26.67	52620	7	2.20	0.087	6.60	0.260	34.28	0.053
2	33.62	66360	7	2.47	0.097	7.42	0.292	43.23	0.067
1	42.41	83690	19	1.69	0.066	8.43	0.332	55.80	0.087
1/0	53.49	105600	19	1.89	0.074	9.45	0.372	70.41	0.109
2/0	67.43	133100	19	2.13	0.084	10.62	0.418	88.74	0.137
3/0	85.01	167800	19	2.39	0.094	11.94	0.470	111.9	0.173
4/0	107.2	211600	19	2.68	0.106	13.41	0.528	141.1	0.219
250	127	—	37	2.09	0.082	14.61	0.575	168	0.260
300	152	—	37	2.29	0.090	16.00	0.630	201	0.312
350	177	—	37	2.47	0.097	17.30	0.681	235	0.364
400	203	—	37	2.64	0.104	18.49	0.728	268	0.416
500	253	—	37	2.95	0.116	20.65	0.813	336	0.519
600	304	—	61	2.52	0.099	22.68	0.893	404	0.626
700	355	—	61	2.72	0.107	24.49	0.964	471	0.730
750	380	—	61	2.82	0.111	25.35	0.998	505	0.782
800	405	—	61	2.91	0.114	26.16	1.030	538	0.834
900	456	—	61	3.09	0.122	27.79	1.094	606	0.940
1000	507	—	61	3.25	0.128	29.26	1.152	673	1.042
1250	633	—	91	2.98	0.117	32.74	1.289	842	1.305
1500	760	—	91	3.26	0.128	35.86	1.412	1011	1.566
1750	887	—	127	2.98	0.117	38.76	1.526	1180	1.829
2000	1013	—	127	3.19	0.126	41.45	1.632	1349	2.092

Notes:
1. These resistance values are valid **only** for the parameters as given. Using conductors having coated strands, different stranding type, and, especially, other temperatures changes the resistance.
2. Formula for temperature change: $R_2 = R_1 [1 + \alpha (T_2 - 75)]$ where $\alpha_{cu} = 0.00323$, $\alpha_{AL} = 0.00330$ at 75°C.
3. Conductors with compact and compressed stranding have about 9 percent and 3 percent, respectively, smaller bare conductor diameters than those shown. See Table 5A for actual compact cable dimensions.

APPENDIX NEC Tables

Direct-Current Resistance at 75°C (167°F)					
Copper				Aluminum	
Uncoated		Coated			
ohm/km	ohm/kFT	ohm/km	ohm/kFT	ohm/km	ohm/kFT
25.5	7.77	26.5	8.08	42.0	12.8
26.1	7.95	27.7	8.45	42.8	13.1
16.0	4.89	16.7	5.08	26.4	8.05
16.4	4.99	17.3	5.29	26.9	8.21
10.1	3.07	10.4	3.19	16.6	5.06
10.3	3.14	10.7	3.26	16.9	5.17
6.34	1.93	6.57	2.01	10.45	3.18
6.50	1.98	6.73	2.05	10.69	3.25
3.984	1.21	4.148	1.26	6.561	2.00
4.070	1.24	4.226	1.29	6.679	2.04
2.506	0.764	2.579	0.786	4.125	1.26
2.551	0.778	2.653	0.809	4.204	1.28
1.608	0.491	1.671	0.510	2.652	0.808
1.010	0.308	1.053	0.321	1.666	0.508
0.802	0.245	0.833	0.254	1.320	0.403
0.634	0.194	0.661	0.201	1.045	0.319
0.505	0.154	0.524	0.160	0.829	0.253
0.399	0.122	0.415	0.127	0.660	0.201
0.3170	0.0967	0.329	0.101	0.523	0.159
0.2512	0.0766	0.2610	0.0797	0.413	0.126
0.1996	0.0608	0.2050	0.0626	0.328	0.100
0.1687	0.0515	0.1753	0.0535	0.2778	0.0847
0.1409	0.0429	0.1463	0.0446	0.2318	0.0707
0.1205	0.0367	0.1252	0.0382	0.1984	0.0605
0.1053	0.0321	0.1084	0.0331	0.1737	0.0529
0.0845	0.0258	0.0869	0.0265	0.1391	0.0424
0.0704	0.0214	0.0732	0.0223	0.1159	0.0353
0.0603	0.0184	0.0622	0.0189	0.0994	0.0303
0.0563	0.0171	0.0579	0.0176	0.0927	0.0282
0.0528	0.0161	0.0544	0.0166	0.0868	0.0265
0.0470	0.0143	0.0481	0.0147	0.0770	0.0235
0.0423	0.0129	0.0434	0.0132	0.0695	0.0212
0.0338	0.0103	0.0347	0.0106	0.0554	0.0169
0.02814	0.00858	0.02814	0.00883	0.0464	0.0141
0.02410	0.00735	0.02410	0.00756	0.0397	0.0121
0.02109	0.00643	0.02109	0.00662	0.0348	0.0106

4. The IACS conductivities used: bare copper = 100%, aluminum = 61%.
5. Class B stranding is listed as well as solid for some sizes. Its overall diameter and area is that of its circumscribing circle.

Source: NEC® Handbook, NFPA, Quincy, MA, 2008, Table 8, Chapter 9

NEC TABLE C.1 Maximum Number of Conductors or Fixture Wires in Electrical Metallic Tubing (EMT)

Type	Conductor Size (AWG kcmil)	Metric Designator (Trade Size)								
		16 (½)	21 (¾)	27 (1)	35 (1¼)	41 (1½)	53 (2)	63 (2½)	78 (3)	91 (3½)	103 (4)
RHH, RHW, RHW-2	14	4	7	11	20	27	46	80	120	157	201
	12	3	6	9	17	23	38	66	100	131	167
	10	2	5	8	13	18	30	53	81	105	135
	8	1	2	4	7	9	16	28	42	55	70
	6	1	1	3	5	8	13	22	34	44	56
	4	1	1	2	4	6	10	17	26	34	44
	3	1	1	1	4	5	9	15	23	30	38
	2	1	1	1	3	4	7	13	20	26	30
	1	0	1	1	1	3	5	9	13	17	22
	1/0	0	1	1	1	2	4	7	11	15	19
	2/0	0	1	1	1	2	4	6	10	13	17
	3/0	0	0	1	1	1	3	5	8	11	14
	4/0	0	0	1	1	1	3	5	7	9	12
	250	0	0	0	1	1	1	3	5	7	9
	300	0	0	0	1	1	1	3	5	6	8
	350	0	0	0	1	1	1	3	4	6	7
	400	0	0	0	1	1	1	2	4	5	7
	500	0	0	0	0	1	1	2	3	4	6
	600	0	0	0	0	1	1	1	3	4	5
	700	0	0	0	0	0	1	1	2	3	4
	750	0	0	0	0	0	1	1	2	3	4
	800	0	0	0	0	0	1	1	2	3	4
	900	0	0	0	0	0	1	1	1	3	3
	1000	0	0	0	0	0	1	1	1	2	3
	1250	0	0	0	0	0	0	1	1	1	2
	1500	0	0	0	0	0	0	1	1	1	1
	1750	0	0	0	0	0	0	1	1	1	1
	2000	0	0	0	0	0	0	1	1	1	1
TW	14	8	15	25	43	58	96	168	254	332	424
	12	6	11	19	33	45	74	129	195	255	326
	10	5	8	14	24	33	55	96	145	190	243
	8	2	5	8	13	18	30	53	81	105	135
RHH*, RHW*, RHW-2*, THHW, THW, THW-2	14	6	10	16	28	39	64	112	169	221	282

		CONDUCTORS								
	Conductor	Metric Designator (Trade Size)								
Type	Size (AWG kcmil)	16 (½)	21 (¾)	27 (1)	35 (1¼)	41 (1½)	53 (2)	63 (2½)	78 (3)	91 (3½)	103 (4)
RHH*, RHW*, RHW-2*, THHW, THW	12	4	8	13	23	31	51	90	136	177	227
	10	3	6	10	18	24	40	70	106	138	177
RHH*, RHW*, RHW-2*, THHW, THW, THW-2	8	1	4	6	10	14	24	42	63	83	106
RHH*, RHW*, RHW-2*, TW, THW, THHW, THW-2	6	1	3	4	8	11	18	32	48	63	81
	4	1	1	3	6	8	13	24	36	47	60
	3	1	1	3	5	7	12	20	31	40	52
	2	1	1	2	4	6	10	17	26	34	44
	1	1	1	1	3	4	7	12	18	24	31
	1/0	0	1	1	2	3	6	10	16	20	26
	2/0	0	1	1	1	3	5	9	13	17	22
	3/0	0	1	1	1	2	4	7	11	15	19
	4/0	0	0	1	1	1	3	6	9	12	16
	250	0	0	1	1	1	3	5	7	10	13
	300	0	0	1	1	1	2	4	6	8	11
	350	0	0	0	1	1	1	4	6	7	10
	400	0	0	0	1	1	1	3	5	7	9
	500	0	0	0	1	1	1	3	4	6	7
	600	0	0	0	1	1	1	2	3	4	6
	700	0	0	0	0	1	1	1	3	4	5
	750	0	0	0	0	1	1	1	3	4	5
	800	0	0	0	0	1	1	1	3	3	5
	900	0	0	0	0	0	1	1	2	3	4
	1000	0	0	0	0	0	1	1	2	3	4
	1250	0	0	0	0	0	1	1	1	2	3
	1500	0	0	0	0	0	1	1	1	1	2
	1750	0	0	0	0	0	0	1	1	1	2
	2000	0	0	0	0	0	0	1	1	1	1

(*continues*)

NEC TABLE C.1 Maximum Number of Conductors or Fixture Wires in Electrical Metallic Tubing (EMT) (*continued*)

Type	Conductor Size (AWG kcmil)	\multicolumn{9}{c|}{CONDUCTORS — Metric Designator (Trade Size)}								
		16 (½)	21 (¾)	27 (1)	35 (1¼)	41 (1½)	53 (2)	63 (2½)	78 (3)	91 (3½)	103 (4)
THHN, THWN, THWN-2	14	12	22	35	61	84	138	241	364	476	608
	12	9	16	26	45	61	101	176	266	347	443
	10	5	10	16	28	38	63	111	167	219	279
	8	3	6	9	16	22	36	64	96	126	161
	6	2	4	7	12	16	26	46	69	91	116
	4	1	2	4	7	10	16	28	43	56	71
	3	1	1	3	6	8	13	24	36	47	60
	2	1	1	3	5	7	11	20	30	40	51
	1	1	1	1	4	5	8	15	22	29	37
	1/0	1	1	1	3	4	7	12	19	25	32
	2/0	0	1	1	2	3	6	10	16	20	26
	3/0	0	1	1	1	3	5	8	13	17	22
	4/0	0	1	1	1	2	4	7	11	14	18
	250	0	0	1	1	1	3	6	9	11	15
	300	0	0	1	1	1	3	5	7	10	13
	350	0	0	1	1	1	2	4	6	9	11
	400	0	0	0	1	1	1	4	6	8	10
	500	0	0	0	1	1	1	3	5	6	8
	600	0	0	0	1	1	1	2	4	5	7
	700	0	0	0	1	1	1	2	3	4	6
	750	0	0	0	0	1	1	1	3	4	5
	800	0	0	0	0	1	1	1	3	4	5
	900	0	0	0	0	1	1	1	3	3	4
	1000	0	0	0	0	1	1	1	2	3	4
FEP, FEPB, PFA, PFAH, TFE	14	12	21	34	60	81	134	234	354	462	590
	12	9	15	25	43	59	98	171	258	337	430
	10	6	11	18	31	42	70	122	185	241	309
	8	3	6	10	18	24	40	70	106	138	177
	6	2	4	7	12	17	28	50	75	98	126
	4	1	3	5	9	12	20	35	53	69	88
	3	1	2	4	7	10	16	29	44	57	73
	2	1	1	3	6	8	13	24	36	47	60
PFA, PFAH, TFE	1	1	1	2	4	6	9	16	25	33	42
PFAH, TFE PFA, PFAH, TFE, Z	1/0	1	1	1	3	5	8	14	21	27	35
	2/0	0	1	1	3	4	6	11	17	22	29
	3/0	0	1	1	2	3	5	9	14	18	24
	4/0	0	1	1	1	2	4	8	11	15	19

Type	Conductor Size (AWG kcmil)	Metric Designator (Trade Size)									
		16 (½)	21 (¾)	27 (1)	35 (1¼)	41 (1½)	53 (2)	63 (2½)	78 (3)	91 (3½)	103 (4)
Z	14	14	25	41	72	98	161	282	426	556	711
	12	10	18	29	51	69	114	200	302	394	504
	10	6	11	18	31	42	70	122	185	241	309
	8	4	7	11	20	27	44	77	117	153	195
	6	3	5	8	14	19	31	54	82	107	137
	4	1	3	5	9	13	21	37	56	74	94
	3	1	2	4	7	9	15	27	41	54	69
	2	1	1	3	6	8	13	22	34	45	57
	1	1	1	2	4	6	10	18	28	36	46
XHH, XHHW, XHHW-2, ZW	14	8	15	25	43	58	96	168	254	332	424
	12	6	11	19	33	45	74	129	195	255	326
	10	5	8	14	24	33	55	96	145	190	243
	8	2	5	8	13	18	30	53	81	105	135
	6	1	3	6	10	14	22	39	60	78	100
	4	1	2	4	7	10	16	28	43	56	72
	3	1	1	3	6	8	14	24	36	48	61
	2	1	1	3	5	7	11	20	31	40	51
XHH, XHHW, XHHW-2	1	1	1	1	4	5	8	15	23	30	38
	1/0	1	1	1	3	4	7	13	19	25	32
	2/0	0	1	1	2	3	6	10	16	21	27
	3/0	0	1	1	1	3	5	9	13	17	22
	4/0	0	1	1	1	2	4	7	11	14	18
	250	0	0	1	1	1	3	6	9	12	15
	300	0	0	1	1	1	3	5	8	10	13
	350	0	0	1	1	1	2	4	7	9	11
	400	0	0	0	1	1	1	4	6	8	10
	500	0	0	0	1	1	1	3	5	6	8
	600	0	0	0	1	1	1	2	4	5	6
	700	0	0	0	0	1	1	2	3	4	6
	750	0	0	0	0	1	1	1	3	4	5
	800	0	0	0	0	1	1	1	3	4	5
	900	0	0	0	0	1	1	1	3	3	4
	1000	0	0	0	0	0	1	1	2	3	4
	1250	0	0	0	0	0	1	1	1	2	3
	1500	0	0	0	0	0	1	1	1	1	3
	1750	0	0	0	0	0	0	1	1	1	2
	2000	0	0	0	0	0	0	1	1	1	1

(*continues*)

NEC TABLE C.1 Maximum Number of Conductors or Fixture Wires in Electrical Metallic Tubing (EMT) (*continued*)

		FIXTURE WIRES					
Type	Conductor Size (AWG/kcmil)	Metric Designator (Trade Size)					
		16 (½)	21 (¾)	27 (1)	35 (1¼)	41 (1½)	53 (2)
FFH-2, RFH-2, RFHH-3	18	8	14	24	41	56	92
	16	7	12	20	34	47	78
SF-2, SFF-2	18	10	18	30	52	71	116
	16	8	15	25	43	58	96
	14	7	12	20	34	47	78
SF-1, SFF-1	18	18	33	53	92	125	206
RFH-1, RFHH-2, TF, TFF, XF, XFF	18	14	24	39	68	92	152
RFHH-2, TF, TFF, XF, XFF	16	11	19	31	55	74	123
XF, XFF	14	8	15	25	43	58	96
TFN, TFFN	18	22	38	63	108	148	244
	16	17	29	48	83	113	186
PF, PFF, PGF, PGFF, PAF, PTF, PTFF, PAFF	18	21	36	59	103	140	231
	16	16	28	46	79	108	179
	14	12	21	34	60	81	134
ZF, ZFF, ZHF, HF, HFF	18	27	47	77	133	181	298
	16	20	35	56	98	133	220
	14	14	25	41	72	98	161
KF-2, KFF-2	18	39	69	111	193	262	433
	16	27	48	78	136	185	305
	14	19	33	54	93	127	209
	12	13	23	37	64	87	144
	10	8	15	25	43	58	96
KF-1, KFF-1	18	46	82	133	230	313	516
	16	33	57	93	161	220	362
	14	22	38	63	108	148	244
	12	14	25	41	72	98	161
	10	9	16	27	47	64	105
XF, XFF	12	4	8	13	23	31	51
	10	3	6	10	18	24	40

Notes:
1. This table is for concentric stranded conductors only. For compact stranded conductors, Table C.1(A) should be used.
2. Two-hour fire-rated RHH cable has ceramifiable insulation which has much larger diameters than other HH wires. Consult manufacturer's conduit fill tables.
*Types RHH, RHW, and RHW-2 without outer covering.

Source: *NEC® Handbook*, NFPA, Quincy, MA, 2008, Table C1

Index

Italicized page locators indicate a figure; tables are noted with a t.

■ A

A/B switching method, 65, 69
AC polyphase motors (other than wound-rotor), percentage of full-load current for, 35t
Allowable Ampacities of Insulated Conductors (NEC Table 310.16), 31
Alternating-current systems
    grounding electrode conductor for, 95t
    proper grounding of, 91
Aluminum, direct-current resistance at 75° C, 33t
Aluminum conductors, voltage drop calculation and DC wire constant value for, 31
American National Standards Institute, 4, 7
American Wire Gauge sizes, 95
    insulation types and, raceway types and conductors of, 39–40
Amperage load, calculating, 3
Amperage size, calculating, for panelboards and transformers, 94
ANSI. *See* American National Standards Institute
Applicable standards, determining, 3–4
Architectural E sheets, E sheets vs., 6
Area category method, 66, 67t, 69
    defined, 69
    energy code regulations met with, 68
AWG size. *See* American Wire Gauge size

■ B

Back voltages, 114
Balanced distribution, 18
    branch-circuit layout designed for, 14
Bid proposal, 128, 131
Blueprints, 4
Bonding jumper, 95
Branch-circuit conductors
    adjusting size of, for voltage drop, 31
    voltage drop and increasing circular mil area of, 44
Branch-circuit distribution
    balancing, 14–15
    raceways for, 38–40
        NEC tables for designing, 38
Branch-circuit distribution section, in single-line diagrams, 89
Branch-circuit identification, 62–63
    for raceways, conductors, and home run detail, 16
Branch-circuit layout, designing for balanced distribution, 14
Branch-circuit numbers, electrical plan review checklist, 129
Branch-circuit raceways, lighting and, 65
Branch-circuit requirements
    determining, 12–16
        balancing branch-circuit distribution, 14–15
        panel schedules, 15–16
Branch circuits, 13, 18, 44
    designating in electrical plan, 16–17
    individual
        common grounded conductor not shared by, 30
        common grounded conductor shared by, 30
        supplied with own ungrounded, separate grounding conductor and individual raceways, 30
    maximum allowable voltage drop for, 31t
    number of, in panelboards, 41
    specialized, for motors, 30
Breaker-based system, distribution and, 88
Building permits, issuance of, 128
Bypass switches
    defined, 69
    programmable controllers and, 64

■ C

CAD. *See* Computer-aided design
Ceiling plans, reflected, 59, 70
CEMFs, 114
Circuit breakers
    handles tied together for, 17
    short-circuit ratings for, 116
Circuitry
    electrical plan review checklist, 129
    individual, 2
    raceways and, electrical symbol list for, 5
Circuits, proper identification for, 18
Coefficient of utilization, 57
Color rendering index, 57, 69
Column multiplier
    maximum mounting distance for lighting fixtures and, 59
Commercial areas, lower luminance levels for, 56
Commercial buildings
    area category method for calculating maximum allowable lighting power in watts for, 68
    specialized lighting design for, 56
Commercial electrical equipment, specialized electrical requirements for, 2
Complete building method, 66, 66t, 69
Computer-aided design, 4, 7
Computers, specialized electrical requirements for, 2
Conductor properties, 32t
Conductors, 17
    branch-circuit identification for, 16
    electrical plan review checklist, 129
    paralleling of, 95–96
    of same size, determining conduit fill for, 41
    values for, 114t
    of variable size, determining conduit fill for, 42
Conduit fill, determining
    for conductors of same size, 41
    for conductors of variable size, 42
Construction phase, beginning, 128, 128
Construction plans
    defined, 131
    finalizing, 128
Construction projects, design plans for, 2
Continuous loads, 13, 18, 91, 97
Control devices, 63–65
    electrical plan review checklist, 129
Copper, uncoated and coated, direct-current resistance at 75° C, 33t
Copper conductors, voltage drop calculation and DC wire constant value for, 31
Copper THWN branch-circuit conductor, sizing, for 15-hp, 208 volt, 3-phase motor, 31
Counter electromotive force, 114, 119
CRI. *See* Color rendering index
Cu. *See* Coefficient of utilization
C value (for conductors), 114, 119

■ D

Demand factor
    calculating, for general purpose receptacle outlets, 111
    defined, 119
    NEC definition of, 110–111
Design B energy-efficient motor, percentage of full-load current for, 35t
Design process, stages in, 2
Digitized plans, 4
Direct current (constant voltage) motor, percentage of full-load current for, 35t
Direct-current resistance at 75° C, 33t
Disconnecting means type and size, electrical plan review checklist, 129
Distribution section, in single-line diagrams, 88–89
Distribution system components
    selecting and sizing, 90–91
        distribution transformers, 91
        panelboards and feeders, 91
Distribution system loads, 111–112
Distribution systems
    conductors used in parallel, 95–96
    defined, 3, 3
    designing electrical distribution equipment, 89–90
    electrical plan review checklist, 129–130
    grounding and, 91, 93–95
        grounding distribution transformers, 94–95
        grounding service entrance equipment, 95
    selecting and sizing components in, 90–91
        distribution transformers, 91
        panelboards and feeders, 91
    single-line diagrams, 86, 87, 88–89
        distribution section of, 88–89
        distribution transformers section of, 89
        panelboard, transformer, and branch-circuit distribution section of, 89
        service entrance conductors, 88
        system grounding methods section of, 89
    sizing for specialized equipment loads, 111
    switchboards, 86
Distribution transformers
    electrical plan review checklist and, 130
    grounding, 94–95
    overcurrent protection for, on primary and secondary sides, 91
Distribution transformers section, in single-line diagrams, 89
Dual-element fuse, 34, 35t, 36

■ E

EASA. *See* Electrical Apparatus and Service Association
Effective grounding path, 91, 97
Electrical Apparatus and Service Association, 4, 7
Electrical distribution equipment, designing, 89–90
Electrical explosions, 110, 110
Electrical metallic tubing, 36
    determining conduit fill for six size 4 AWG THWN conductors to be installed in, 41
Electrical Plan Review Checklist, 128–130, 134–135
    distribution system, 129–130
    general and specialized electrical requirements, 129
    lighting system, 129
Electrical plan review process, stages in, 128
Electrical plans
    branch circuits designated in, 16–17
    creating, 4, 6
    defining parts of, 2
    identifying, 6
Electrical requirements
    general, 2
    specialized, 2
        Electrical symbol list, 5, 9
Electrical symbols
    creating new, 6
    referencing on electrical plan, 4, 7
Energy code design methods, understanding, 66–67
Energy code requirements
    complying with, 66
    meeting, with area category method, 68
Energy management, 67
Energy management controls, electrical plan review checklist, 129
Energy-savings requirements, lighting energy codes and, 65
Energy-saving technologies, lighting system requirements and, 3
Equipment branch circuits, panel schedules and, 42–43
Equipment grounding conductors, 28
    defined, 44
Equipment grounding methods, determining, 30
Equipment lists, 44
    defined, 44
    information on, 28, 29
Errors, electrical plan review and checking for, 128, 131
E sheets, presentation of, 6
Even-numbered circuits, in panel schedules, 15

■ F

Fc. *See* Foot candle
Feeders, panelboards and, 91
Fixed multioutlet assemblies, requirements of 220.14 (h)(I) and (H)2 as applied to, 14
Fixture lumens, 57

F

Fixture quantity, 61
Fixtures, electrical symbol list for, 5
FL. *See* Fixture lumens
Flexible metallic conduit, 28
Fluorescent lamps, lumens per watt ratios for, 57
Focused light levels, 59
Foot candle, 56, 70
Full Load Current, Three Phase Alternating Current Motors (*NEC* Table 430.250), 31
Full Load Currents in Amperes, Direct Current Motors (*NEC* Table 430.247), 31
Full Load Currents in Amperes, Single Phase Alternating Current Motors (*NEC* Table 430.248), 31
Fuse-based system, distribution and, 88
Fuses, short-circuit amperages and, 117

G

General electrical loads, 110–111
General electrical requirements, 2
  electrical plan review checklist, 129
General purpose branch circuit, 13, 19
General purpose 120-volt receptacles, to be served by 20-A branch circuit, determining number of, 13
General purpose receptacle outlet, office space requirements for, 12, *12*
General purpose receptacles
  branch-circuit identification for, *16*
  spacing for, *12*
General purpose requirements, determining, 12
Ground-fault protection devices
  motor branch-circuit short-circuit and, 34
  NEC Table 430.52 maximum rating or setting of, 35*t*
  sizing, 37
Ground faults
  branch circuit conductors and protection from, 34, 44
  defined, 44
Grounding
  illustration of, *95*
  methods, 17
  special, 2
  of motors and equipment, 36–37
  understanding, 91, 93
Grounding conductors, adjusting size of equipment of, for voltage drop, 39
Grounding electrode, 95, 97
Grounding electrode conductor
  defined, 97
  sizing, 95, 96
Grounding path, effective, 91

H

Halogen lighting fixtures, 57
High-density discharge (HID), 74
Home run detail, branch-circuit identification for, 16, *16*
Home run raceway designation, electrical plan review checklist, 129
Home runs, 16, 17, 19

I

IESNA. *See* Illuminating and Engineering Society of North America
IESNA Lighting Handbook, 56
IG. *See* Isolated ground
Illuminating and Engineering Society of North America, 4, 7, 56
Incandescent lamps, lumens per watt ratios for, 57
Inductance to resistance ratio (X/R), calculating, 114
Input, luminance and, 57
Inrush currents, 31, 44
Instantaneous trip breaker, 34, 35*t*, *36*
Inverse time breaker, 34, 35*t*, *36*
Isolated ground, 30, 45

K

kcmil sizes and insulation types, raceway types and conductors of, 39–40
Keynotes sections, 65–66

L

Lamp lumens, 58*t*
Larger motors, amperage capacity of panelboards and, 42
LED. *See* Light emitting diode
Light distribution, fixtures and mounting heights for, *60*
Light-emitting diode, 61
Lighting branch circuits, 62
  defined, 70
  determining maximum volt-ampere rating for, 64
  maximum allowable volt-ampere rating for, 63
  number required, 63, 64
Lighting design, goal of, 56
Lighting design procedures
  area category method, 67*t*
  complete building method, 66*t*
Lighting energy codes, energy-savings requirements dictated by, 65
Lighting energy-saving devices, 67, 70
Lighting fixtures, 69
  calculating required number of, 57–58
  determining location for, 59, 61
  electrical plan review checklist, 129
  identification method for, 63
  quality of, 58
  selecting, 57, 69
Lighting fixture schedule, 59, 61
  defined, 70
  details listed on, 62
  electrical plan review checklist, 129
  information given by, 59, 61
Lighting plans, 69
  creation of, 59–65
    branch-circuit identification, 62–63
    control devices, 63–65
    lighting branch-circuit raceways, 65
    lighting fixture schedule, 59, 61
  elements within, 59
Lighting requirements, determining, 56
Lighting systems, 3, *3*
  components in, 56
  defined, 70
  designing panelboards for, *92*, 92–93
  electrical plan review checklist, 129
  energy code design methods and, 66–68
  energy code requirements and, 66
  keynotes sections and, 65–66
  lighting fixtures
    calculating number required, 57–58
    location for, 59, 61
    selecting, 57
  lighting plan creation, 59, 61–65
    branch-circuit identification, 62–63
    control devices, 63–65
    lighting branch-circuit raceways, 65
    lighting fixture schedule, 59, 61, 62
  lighting requirements, 56
  loads, 111
  national or state-mandated energy-saving requirements for, 4
Lighting whiteline levels, recommended, 56*t*
Light loss factor (LLF), 57, 78
lm/W. *See* Lumens per watt
Load amperage, electrical plan review checklist and, 130
Load calculations
  defined, 119
  determining for motor loads, 112
  electrical plan review checklist, 129
  performing, 110–112
    distribution system loads, 111–112
    general electrical loads, 110–111
    lighting system loads, 111
    specialized electrical loads, 111
  purpose of, 110
Loads, continuous, 91, 97
Load values, distribution systems and, 4
Lumens, lamp, 58*t*
Lumens per watt, 57, 70
Luminance, 57
Luminance levels, 56, *56*, 69
  defined, 70
  proper vs. improper design and, *60*
  for showrooms, 77

M

Machinery and equipment, specialized electrical requirements for, 2
Magnetic principles, grounding distribution transformers and, 94–95
Main disconnecting means, 86, 97
Main switchboards, 3, 86, *86*
Manufacturer electrical specification sheets, 28, 45
Manufacturing areas
  high-level lighting in, 57
  lighting whiteline levels recommended for, 56*t*
Maximum conduit fill for installation, determining, 40
Metal halide lighting fixtures, 57
Microwaves
  specialized electrical requirements for, 2
Minimum Size Equipment Grounding Conductors for Grounding Raceway and Equipment (*NEC* Table 250.122), 38*t*
Motor branch-circuit overload devices, 34, 36
Motor branch-circuit short-circuit
  ground-fault protection devices and, 34
  NEC Table 430.52 maximum rating or setting of, 35*t*
Motor controller, motor overload device in, 38
Motor equipment branch-circuit conductors
  sizing, 31
    motor voltage and full-load amperage relative to, 30
    NEC tables for, 31
Motor loads
  load calculations for, 112
Motor overload device, in motor controller, 38
Motors, 30–31, 34–38
  grounding of, 36–37
  large, determining number of panelboards and, 42
  NEC guidelines for, 30
Mounting distance, for lighting fixtures, row and column multipliers and, 59
MR16 lamps, 78
Multiwire branch circuits, 14, *14*, 19

N

National Electrical Code ("the *Code*"), 3, 7, 9
  branch-circuit requirements and, 13
  general purpose receptacles and, 12
  lighting branch circuits requirements and, 63
  lighting system requirements and, 3, *3*
  service entrance conductors defined by, 88
National Electrical Manufacturers Association, 4, 7, 9
  standard panelboard and disconnect sizes, 91, 91*t*
National Fire Protection Association, 3
NEC. *See National Electrical Code*
NEMA. *See* National Electrical Manufacturers Association
Network servers, specialized electrical requirements for, 2
NFPA. *See* National Fire Protection Association
Non-time delay fuse, 34, 35*t*, 36

O

Occupancy sensors, 63–64, 67, 70
Odd-numbered circuits, in panel schedules, 15
Office areas
  lighting whiteline levels recommended for, 56*t*

INDEX

Office space illumination levels, complete building method and, 66
Ohm's law
  designing panelboards for lighting systems and, 92
  determining full-load amperage for transformer and use of, 115
  for parallel circuits, 96
  total volt-ampere rating calculated with, 43
120/208-volt, 3-phase, 4-wire, wye system, 14, 18
120-volt general purpose receptacle outlets, 2
Outlets, electrical symbol list for, 5
One-line diagram, 86. *See also* Single–line diagrams
Overcurrent protection sizes, primary and secondary, electrical plan review checklist and, 130
Overcurrent protective devices, 86
  determining size of, 30
Overload, 13, 19
  avoiding, specialized equipment branch circuits and, 28

■ P

Panelboards, 3
  assignment of, 90–91
  calculating amperage size of, 94
  designing, 44
    for lighting systems, 92, 92–93
  determining available short-circuit current for, 117
  determining number of, 40–43
    equipment branch circuits and panel schedules, 42–43
    larger motors, 42
    minimum, 42
  feeders and, 91
  number of branch-circuit spaces in, 41
  service capabilities with, 90
Panelboard section, in single-line diagrams, 89
Panel schedules, 15–16, 19
  design goals and, 18
  electrical plan review checklist, 129
  equipment branch circuits and, 42–43
  120/208-volt, 3-phase, 4-wire, 15
Parallel circuits, Ohm's law for, 96
Photocopiers, specialized electrical requirements for, 2
Plan check engineer, 128, 131
Point-to-point method
  defined, 119
  short-circuit calculations and, 114, 115–116
Power panelboards, designations for, 89
Primary grounding electrode, 94–95
Programmable lighting controllers, 64, 70
Project scope, understanding, 2
Protective devices, 36
  types of, 34
PVC, cost reduction and use of, 39

■ R

Raceway and equipment, grounding, *NEC* Table 250.122 minimum size equipment grounding conductors for, 38t
Raceway legend, 44
  defined, 45
  electrical plan review checklist, 129
  information provided by, 43, 43
Raceway methods
  selecting and sizing, 30
  variance in, 17
Raceways, 16
  for branch-circuit distribution, 38–40
  branch-circuit identification for, 16
  calculating proper size of, 39
  defined, 19
  electrical plan review checklist, 129
    single, larger, with multiple branch circuits for home run, 40

Receptacle outlets, 12, 19
Receptacles
  electrical plan review checklist, 129
  value of, per *NEC*, 13
Reflected ceiling plans, 59, 70
Remodeling building projects, building plans and, 2
Rigid nonmetallic conduits
  cost reduction and use of, 39
Row multiplier
  maximum mounting distance for lighting fixtures and, 59
Running current, size of, *vs.* size of starting current, 31

■ S

Scope of project, understanding, 2, 7, 9
Screen glare, lighting and reduction of, 56
Sensors, occupancy, 63–64, 67, 70
Separately derived system, 94, 97
Service and equipment, electrical symbol list for, 5
Service entrance conductors, 86
  defined, 97
  electrical plan review checklist, 129
  installing, underground and overhead or aboveground, 88
Service entrance equipment, grounding of, 95
Short-circuit amperages
  calculations for, on electrical plans, 114
  eliminating hazards associated with, 114, 117–118, 119
Short-circuit calculations
  defined, 119
  electrical plan review checklist, 129
  performing, 112–118
    transformer impedance, 113–114
  purpose of, 112
Short-circuit current
  available, determining for panelboard, 117
  determining for transformer, 113
  point-to-point method and determination of, 114, 115–116
Short-circuit devices, sizing, 37
  Short-circuit ratings, for circuit breakers, 116
Short circuits
  branch circuit conductors and protection from, 34, 44
  defined, 45
Short-circuit values, calculating, 3
Single-line diagrams, 97
  defined, 97, 131
  distribution section in, 88–89
  distribution transformers section in, 89
  electrical plan review checklist and, 130
  electrical plan review process and, 128
  format of, 88
  information in, 86, 87
  panelboard, transformer, and branch-circuit distributions section in, 89
  selecting/sizing distribution system components, 90
  service entrance conductors, 88
  system grounding methods section in, 89
  understanding, 86, 88–89
Single-phase motors, percentage of full-load current for, 35t
Single-phase systems, calculating correct wire size to account for voltage drop in branch circuits, 31
Sizing, short-circuit and fault-protection devices, 37
Spacing criteria, defined, 70
Spacing criteria data, for lighting fixtures, 59
Specialized electrical loads, 111
Specialized electrical requirements, 2
  electrical plan review checklist, 129
Specialized equipment branch circuits, 28, 30
Specialized equipment branch circuit wiring methods, types of, 28, 30

Split duplex receptacles, 17
Squirrel cage (other than Design B energy-efficient), percentage of full-load current for, 35t
Standards, determining, 3–4, 7
Starting current, size of, *vs.* size of running current, 31
Step-down transformers, 63, 70
Switchboards
  defined, 86, 97
  main, 3, 86, 86
Switchboard size, electrical plan review checklist, 129
Switches, 69
  electrical symbol list for, 5
Synchronous motor, percentage of full-load current for, 35t
System grounding electrode conductor, electrical plan review checklist, 129
System grounding methods section, in single-line diagrams, 89

■ T

10 percent load balancing provision, 129
3-phase systems, calculating correct wire size to account for voltage drop in branch circuits, 31
Three-phase transformers, full load currents, 93t
Time clocks, 67
Track lighting fixtures, calculating for, 78
Transformer impedance, 113, 117–118
Transformer impedance rating, 113, 119
Transformers
  calculating amperage size of, 94
  short-circuit current determined for, 113
Transformer section, in single-line diagrams, 89
277/480-volt, 3-phase, 4-wire, wye system, defined, 44
Type EMT conduit, 36
  determining conduit fill for, 42
  determining conduit fill for six size 4 AWG THWN conductors to be installed in, 41
Type THWN copper conductors, 93

■ U

Underground raceways, 24
U.S. Department of Energy, 66
Utility metering equipment, 86, 97

■ V

Vending machines
  specialized electrical requirements for, 2
Voltage drop, 31
  adjusting size of branch-circuit conductors for, 35
  adjusting size of equipment grounding conductors for, 39
  in branch circuits, calculating correct wire size to account for, 31
  defined, 45
  increasing circular mil area of branch circuit conductors and, 44
  indicating on single-line diagrams, 89
  maximum, for branch circuits, 31t
Voltage surges, motorized equipment loads and, 90
Volt-ampere ratings
  for lighting branch circuits, determining, 64
  maximum allowable, for lighting branch circuits, 63
  total, Ohm's law and, 43

■ W

Watts per square foot (W/ft^2), 66
Wire size, correct, calculating to account for voltage drop in branch circuits, 31
Wound-rotor motor, percentage of full-load current for, 35t
Wye, 18

Credits

Chapter 1

Chapter Opener © Ace Stock Limited/Alamy Images; **1-1** © Zdravko Bajazek/Dreamstime.com; **1-2** Courtesy of Duane Romanell; **1-3** Courtesy of David Dooling; **1-4** © Corbis Premium RF/Alamy Images

Chapter 2

Chapter Opener © AbleStock; **2-1** © Chad McDermott/ShutterStock, Inc.; **2-3** NFPA 70®, *National Electrical Code®* and *NEC®* are registered trademarks of the National Fire Protection Association, Quincy, MA.

Chapter 3

Chapter Opener © Fotomy/Dreamstime.com; **3-T2** Source: *NEC® Handbook*, NFPA, Quincy, MA, 2008, Table 8; **3-T3** Source: *NEC® Handbook*, NFPA, Quincy, MA, 2008, Table 430.52; **3-5A**, **3-5B** Courtesy of Cooper Bussmann; **3-5C** Courtesy of General Electric Company, **3-5D** Courtesy of Siemens Industry, Inc.; **3-6** Courtesy of General Electric Company; **3-T4** Source: *NEC® Handbook,* NFPA, Quincy, MA, 2008, Table 250.122; **Page 41** Source: *NEC® Handbook*, NFPA, Quincy, MA, 2008, Table C1

Chapter 4

Chapter Opener © jkitan/ShutterStock, Inc.; **4-1** © Cedric Crucke/ShutterStock, Inc.; **4-2** © jkitan/Dreamstime.com

Chapter 5

Chapter Opener © Jones and Bartlett Publishers. Photographed by Kimberly Potvin; **Page 95** Source: *NEC® Handbook*, NFPA, Quincy, MA, 2008, Table 250.66

Chapter 6

Chapter Opener © Lisa F. Young/ShutterStock, Inc.; **6-1** COPYRIGHT 2006 COASTAL TRAINING TECHNOLOGIES CORPORATION; **6-2** Courtesy of General Electric Company

Chapter 7

7-1 © Corbis Premim RF/Alamy Images

Appendix

Page 144 (Top) Source: *NEC® Handbook*, NFPA, Quincy, MA, 2008, Table 220.44; **Page 144 (Bottom)** Source: *NEC® Handbook,* NFPA, Quincy, MA, 2008, Table 250.122; **Page 145** Source: *NEC® Handbook*, NFPA, Quincy, MA, 2008, Table 250.66; **Page 146** Source: *NEC® Handbook*, NFPA, Quincy, MA, 2008, Table 310.16; **Page 148** Source: *NEC® Handbook*, NFPA, Quincy, MA, 2008, Table 430.247; **Page 149** Source: *NEC® Handbook*, NFPA, Quincy, MA, 2008, Table 430.248; **Page 150** Source: *NEC® Handbook*, NFPA, Quincy, MA, 2008, Table 430.250; **Page 151 (Top)** Source: *NEC® Handbook*, NFPA, Quincy, MA, 2008, Table 430.52; **Page 151 (Bottom)** Source: *NEC® Handbook,* NFPA, Quincy, MA, 2008, Table 430.3(B); **Page 152** Source: *NEC® Handbook*, NFPA, Quincy, MA, 2008, Table 4, Chapter 9; **Page 153** Source: *NEC® Handbook*, NFPA, Quincy, MA, 2008, Table 5, Chapter 9; **Page 154** Source: *NEC® Handbook*, NFPA, Quincy, MA, 2008, Table 8, Chapter 9; **Page 156** Source: *NEC® Handbook*, NFPA, Quincy, MA, 2008, Table C1

Unless otherwise indicated, all tables and illustrations have been provided by John Hauck, and are under copyright of Jones and Bartlett Publishers, LLC.

Reprinted with permission from NFPA 70HB-2008, *National Electric Code® Handbook*, Copyright © 2008, National Fire Protection Association, Quincy, MA 02169. This reprinted material is not the complete and official position of the NFPA on the referenced subject, which is represented only by the standard in its entirety.